# Document Processing Using Machine Learning

# Document Processing Using Machine Learning

Edited by
Sk Md Obaidullah, KC Santosh,
Teresa Gonçalves, Nibaran Das
and Kaushik Roy

CRC Press
Taylor & Francis Group
Boca Raton  London  New York

CRC Press is an imprint of the
Taylor & Francis Group, an **informa** business

A CHAPMAN & HALL BOOK

CRC Press
Taylor & Francis Group
52 Vanderbilt Avenue,
New York, NY 10017

© 2020 by Taylor & Francis Group, LLC
CRC Press is an imprint of Taylor & Francis Group, an Informa business

No claim to original U.S. Government works

Printed on acid-free paper

International Standard Book Number-13: 978-0-367-21847-8 (Hardback)

**Visit the Taylor & Francis Web site at**
**http://www.taylorandfrancis.com**

**and the CRC Press Web site at**
**http://www.crcpress.com**

# Contents

# *Preface*

We are surrounded by huge volumes of data in various categories. In the context of big data, automated and faster tools are in great demand. Often, data exist on scanned documents and can be either machine-printed or handwritten. These include, but are not limited to, letters, checks, payment slips, income tax forms and business forms numbering in the billions. Manual digitization is impossible, time consuming, expensive and vulnerable to errors. Automation can solve such problems. Compared to machine-printed documents, handwritten document processing can be more challenging and, of course, opens a wide range of issues to consider. Writer identification via handwriting can be one of the important applications. Further, we have experienced that both machine-printed and handwritten texts can go together in one document. Can we treat/process them separately? For this, one needs to be able to extract meaningful information; it can be from either machine-printed or handwritten texts. This can open up ideas of document information extraction and understanding, such as information retrieval and natural language processing.

Advances in document processing can only be possible by considering appropriate machine learning algorithms that help build data-driven models or predictions. The process includes classification, clustering, regression, association rules and many more elements. How can we forget deep learning-based approaches? In this book, advanced document processing techniques using up-to-date and well-trained machine learning algorithms are presented.

Chapter 1 discusses the role of AI for different document analysis problems, such as optical character recognition and web text categorization, where real-world issues are considered.

Chapter 2 discusses multiple methods for optical character recognition on handwritten isolated Bangla characters. It provides results on a database of 89,000 plus handwritten characters in addition to two publicly available benchmark databases, namely the ISI handwritten character database and CMATERdb 3.1.2.

Chapter 3 presents a script identification study on multi-script artistic documents where real-world problems are taken into consideration. The primary motivation behind this work is that OCR tools are script-dependent. In the authors' work, a semi-automated segmentation algorithm has been used for character separation within words followed by a thinning procedure and a structural feature; Gabor filters are used for feature extraction. The results were reported using several different machine learning classifiers.

Chapter 4 discusses the use of extreme learning machines (ELMs), as they suppress the pitfalls of neural networks like slow learning rates and parameter tuning. The authors provide an idea of how document analysis can be done using an ELM.

Internet technology has brought about substantial changes to our day-to-day lives. Nowadays, we need several digital means to automatically manage, filter, store and retrieve information from text documents. Automated text document analysis and mining are becoming an essential part of computer applications and thus various classification and clustering approaches are required for carrying out these tasks. The classification of documents needs to be performed in the training dataset, which is further used to train the model classifier to classify the text documents into their respective text categories or domains. Thus, text analysis becomes one of the major aspects of text data mining. Under this scope, Chapters 5, 6 and 7 provide various graph-based models for different web text classification problems.

Chapter 8 discusses two essential aspects of an author, namely author aggression and author profile. The identification of aggression is classified into three classes: overtly, covertly and non-aggressive. On the other hand, profiling identifies the two properties of age and gender. The chapter also explains various machine learning concepts incorporated into the deep neural network. It also provides insight into the previous work done on author aggression and profiling.

Document image analysis problems broadly fall under two categories: offline and online. In Chapter 9, the authors discuss online handwritten Bangla character recognition using features computed from generated offline images. Interestingly, this chapter shows an integrated version/concept of how offline and online domains can be merged to advance recognition performance.

Chapter 10 discusses handwritten character recognition for palm-leaf manuscripts. Palm-leaf manuscripts are ancient documents primarily from South and Southeast Asia. Over hundreds of years, the manuscripts have become damaged. The chapter discusses how to transform the palm-leaf images into machine-encoded texts so that the document can be edited, retrieved, accessed and processed. The issue of applying optical handwritten character recognition for the said problem, with associated challenges, is covered in detail.

We hope this book will help researchers in document analysis and machine learning as different real-life problems are discussed with experimental outcomes. Additionally, undergraduate or postgraduate scholars who wish to carry out their research-based projects or theses on document analysis problems can get help from this book.

**Sk Md Obaidullah**
**KC Santosh**
**Teresa Gonçalves**
**Nibaran Das**
**Kaushik Roy**

# *Editors*

**Sk Md Obaidullah** earned a PhD (Engg.) from Jadavpur University, an MTech in Computer Science and Application from the University of Calcutta and a BE in Computer Science and Engineering from Vidyasagar University in 2017, 2009 and 2004, respectively. He was an Erasmus post-doctoral fellow funded by the European Commission at the University of Évora, Portugal, from November 2017 to September 2018. He has more than 11 years of professional experience including two years in industry and nine years in academia, out of which five years were spent on research. Presently, he is working as an Associate Professor in the Department of Computer Science and Engineering, Aliah University, Kolkata. He has published more than 65 research articles in renowned journals and reputed national/international conferences. He is an active researcher in the fields of document image processing, medical image analysis, pattern recognition and machine learning.

**KC Santosh** (Senior Member, IEEE) is an Assistant Professor and Graduate Program Director for the Department of Computer Science at the University of South Dakota (USD). Dr. Santosh also serves the School of Computing and IT, Taylor's University, as a Visiting Associate Professor. Before joining USD, Dr. Santosh worked as a research fellow at the U.S. National Library of Medicine (NLM), National Institutes of Health (NIH). He worked as a postdoctoral research scientist at the LORIA research centre, Université de Lorraine, in direct collaboration with industrial partner ITESOFT in France. He also worked as a research scientist at the INRIA Nancy Grand Est research centre, France, where, he completed his PhD in Computer Science. Before that, he worked as a graduate research scholar at SIIT, Thammasat University, Thailand. He has published more than 120 peer-reviewed research articles, authored two books (Springer) and edited ten books, journal issues and conference proceedings. Dr. Santosh serves as an associate editor for the *International Journal of Machine Learning & Cybernetics*. Dr. Santosh demonstrates expertise in artificial intelligence, machine learning, pattern recognition, computer vision, image processing, data mining and big data with various application domains, such as healthcare and medical imaging, document information content exploitation, biometrics, forensics, speech/audio analysis, satellite imaging, robotics and the Internet of Things.

**Teresa Gonçalves** is an Assistant Professor in the Department of Informatics at the University of Évora, Portugal. She has a PhD in Informatics from the University of Évora (2008) and a five-year degree and master in Informatics Engineering, both from the Faculty of Sciences and Tecnology, New University of Lisbon, in 1992 and 1996, respectively.

She has published more than 60 research papers in reputed journals and conferences and has worked as the organizing and programme committee chair of various international conferences. She has worked as a PI for different research and mobility projects funded by the Portuguese government and European Commission. Her research interests include machine learning and data mining, namely with textual data and images, recommendation systems and evolutionary algorithms. She is responsible for several courses of undergraduate, masters and doctorate levels in Computer Science. Having successfully supervised two doctorate and six master students, she currently supervises six PhD and six master students, mainly on applying and adapting machine learning approaches to text or image-related problems.

**Nibaran Das** is an Associate Professor in the Department of Computer Science and Engineering at Jadavpur University. Before joining Jadavpur University, from 2005 to 2006, Dr. Das worked as a lecturer in Techno India, Saltlake. He worked as a postdoctoral research scientist at the University of Évora for six months, non-consecutively, between 2012 and 2014. He also worked as a research intern at the Competence Center Multimedia Analysis and Data Mining (MADM) at the DFKI, University of Kaiserslautern, Germany, in 2007. Dr. Das serves as an associate editor for the journal *Sadhana: Academy Proceedings in Engineering Sciences*. Dr. Das has demonstrated expertise in deep learning; pattern recognition; image processing and machine learning with various applications in handwriting recognition, especially character recognition; and medical image analysis. He has published more than 125 research articles, as well as the book *Handbook of Research on Recent Developments in Intelligent Communication Application* (IGI Global) and has co-authored several conference proceedings. He has guided more than 30 master's degree students in his department. Dr. Das also served as a chairperson of the young professional affinity group, IEEE Kolkata section, from 2014–2015. He is the founding editor of the Bangla monthly computer magazine *Computer Jagat*. He is a regular reviewer for high-quality journals (IEEE, Springer, Elsevier) and high-quality conferences and workshops (sponsored by IEEE and Springer) in his domains of expertise.

 **Kaushik Roy** earned a BE in Computer Science & Engineering from NIT Silchar, and an ME and PhD (Engg.) in Computer Science and Engineering from Jadavpur University in 1998, 2002 and 2008, respectively. He has worked as a project-linked personnel in ISI-Kolkata and as a Scientific Officer in CDAC-Kolkata. He has also worked as an Assistant Professor in Maulana Abul Kalam Azad University of Technology, India, formerly known as West Bengal University of Technology. He is currently working as a Professor and Head of the Department of Computer Science, West Bengal State University, Kolkata, India. In 2004, he received the Young IT Professional award from the Computer Society of India. He has published more than 150 research papers/book chapters in reputed conferences and journals. His research interests includes pattern recognition, document image processing, medical image analysis, online handwriting recognition, speech recognition and audio signal processing. He is a life member of IUPRAI (a unit of IAPR) and the Computer Society of India.

# Contributors

**Roy Bayot**
University of Évora
Évora, Portugal

**Ankan Bhattacharyya**
Future Institute of Engineering &
    Management
Kolkata, India

**Jakramate Bootkrajang**
Chiang Mai University
Chiang Mai, Thailand

**Jeerayut Chaijaruwanich**
Chiang Mai University
Chiang Mai, Thailand

**Sahana Das**
West Bengal State University
Kolkata, India

**Niladri Sekhar Dash**
Indian Statistical Institute
Kolkata, India

**Ankita Dhar**
West Bengal State University
Barasat, Kolkata, India

**Mridul Ghosh**
Shyampur Siddheswari
    Mahavidyalaya
Howrah, India

**Subhashmita Ghosh**
Aliah University
Kolkata, India

**Chayan Halder**
University of Engineering and
    Management
Kolkata, India

**Papangkorn Inkeaw**
Chiang Mai University
Chiang Mai, Thailand

**Himadri Mukherjee**
West Bengal State University
Kolkata, India

**Santanu Phadikar**
Maulana Abul Kalam Azad
    University of Technology
Kalyani, India

**Kashyap Raiyani**
University of Évora
Évora, Portugal

**Payel Rakshit**
Maheshtala College
Kolkata, India

**Shibaprasad Sen**
Future Institute of Engineering &
    Management
Kolkata, India

# 1

# Artificial Intelligence for Document Image Analysis

**Himadri Mukherjee, Payel Rakshit, Ankita Dhar,
Sk Md Obaidullah, KC Santosh, Santanu Phadikar
and Kaushik Roy**

## CONTENTS

## 1.1 Introduction

There has been rapid development in technology which has aided in the digitization of documents. The number of digital documents has increased significantly over time [1, 2]. Information is now easily available on the Internet and can be distributed with ease. Such voluminous numbers of documents demand efficient processing. Digitized documents can be broadly categorized into two types, namely document images and text documents. In the case of document images, it is first essential to understand what is written. This requires optical character recognition (OCR) [3–5]. Once the characters are identified, approaches based on natural language processing (NLP) [6–8] need to be used to understand what is written. In the case of text documents, research in the fields of OCR and NLP started way back in the last century and different systems in languages like English are now commercially available [9–11], but there have not been significant developments for Indic languages. One reason for this is the complex nature of Indic scripts [12]. This is also coupled with the unavailability of standard (and free) datasets for research.

## 1.2 Optical Character Recognition

Optical character recognition [13, 14] refers to the task of decoding what is written in a document. It does not involve understanding the written texts, but it does involve converting a scan or a picture of a document, identifying the characters and giving the text output of the identified words and characters. The document can be either handwritten or printed. In the case of handwritten documents, there are different variations which need to be considered prior to recognition. While writing, it is often observed that the texts have disparate degrees of slants. It is very important to deal with such slants while processing the documents. A document with characters having multiple degrees of slant is presented in Figure 1.1.

The second important factor which needs to be tackled is the similarity between different characters. For instance, the numeral "3" is similar to "ও"

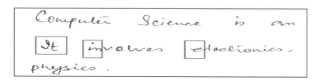

**FIGURE 1.1**
A document depicting multiple degrees of slant for different characters.

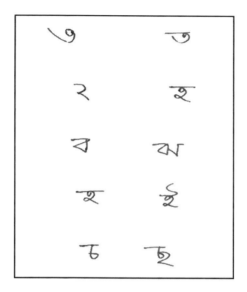

**FIGURE 1.2**
Similarity between different characters in Bangla.

in Bangla. This is illustrated in Figure 1.2. It is very important to handle such cases or else, if interpreted wrongly, the entire sentence might change.

Another important aspect is inter-writer and intra-writer variation. It is often observed that handwritten texts show variation at the character level. That is, the same character is slightly different when written in two instances. This is known as intra-writer variation. Another variation is observed when two different writers write the same thing. The handwriting of disparate people differ from each other in most cases. This is known as inter-writer variation. Thus the system should be able to handle such differences. Inter- and intra-writer variations for a Bangla text are presented in Figures 1.3 and 1.4.

There are 11 official Indic scripts which are used to write in various languages like Bangla, Hindi, Telugu, Tamil, Urdu, etc. The details of the scripts in the languages in which they are used, the approximate total population using each one and the areas in India where they are used are presented in Table 1.1. These scripts have a certain degree of similarity among themselves [15] and thus the OCR system must be able to detect fine disparities to produce accurate results.

## 1.2.1 Dealing with Noise

Any kind of unwanted data is known as noise. It is a problem-specific term. A particular kind of data may be regarded as noise for a certain problem while being of prime importance in a different problem. Real life data is not very pretty. It is often contaminated with disparate types of noises. Handwritten documents are no exception to this and they are often degraded and noisy.

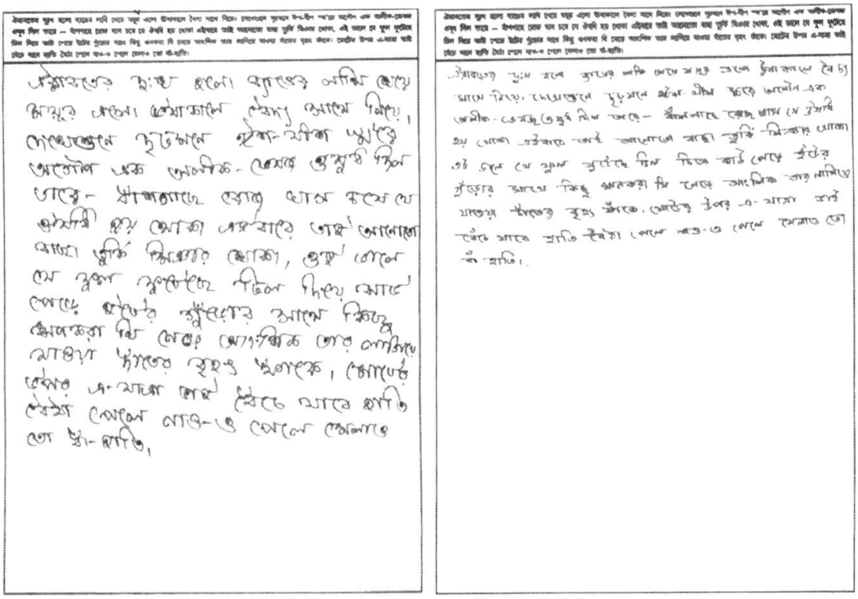

**FIGURE 1.3**
A piece of text written by two different writers.

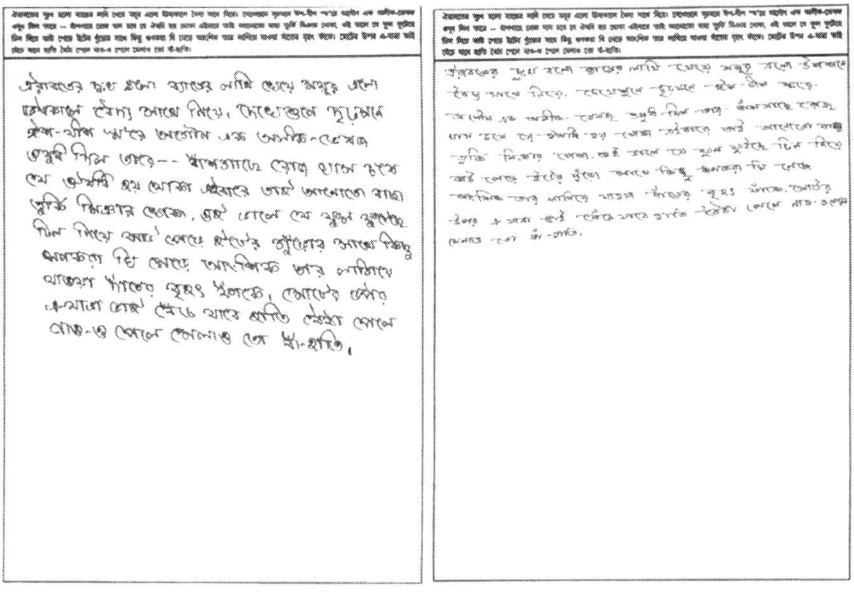

**FIGURE 1.4**
The same text written by the same writers as in Figure 1.3. This is presented to show intra-writer variation.

**TABLE 1.1**

Details of the Indic Scripts

| Language | Script Used | Approximate Number of Users (Millions) |
|---|---|---|
| Hindi | Devanagari | 182.00 |
| Marathi | Devanagari | 68.10 |
| Konkani | Devanagari | 7.60 |
| Sanskrit | Devanagari | 0.03 |
| Sindhi | Devanagari | 21.40 |
| Nepali | Devanagari | 13.90 |
| Maithili | Devanagari | 34.70 |
| Bodo | Devanagari | 0.50 |
| Bangla | Bangla | 181.00 |
| Assamese | Bangla | 16.80 |
| Manipuri | Bangla | 13.70 |
| Telugu | Telugu | 69.80 |
| Tamil | Tamil | 65.70 |
| Urdu | Urdu | 60.60 |
| Kashmiri | Urdu | 5.60 |
| Gujarati | Gujarati | 46.50 |
| Malayalam | Malayalam | 35.90 |
| Oriya | Oriya | 31.70 |
| Kannada | Kannada | 3.63 |
| Punjabi | Gurumukhi | 1.05 |
| Dogri | Gurumukhi | 3.80 |
| Santali | Roman | 6.20 |

Some of these noises include unwanted pen strokes, ink spills, paper creases, stains, etc., which are illustrated in Figures 1.5 and 1.6.

Sometimes the paper may also get stained due to different factors. Another source of noise is scribbles. People often make mistakes while writing and real-life scenarios do not have an undo option. Upon making a mistake, people either scribble it out or just draw a line over the section. It is very

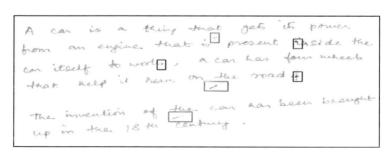

**FIGURE 1.5**

Ink spills (black box) and accidental marks (gray box) in a document.

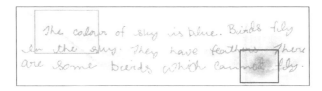

**FIGURE 1.6**
Paper fold (light gray box) and dirt marks (dark gray box) in a document.

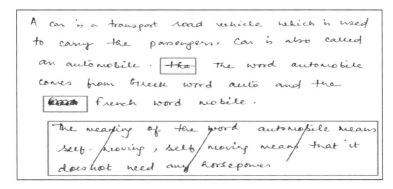

**FIGURE 1.7**
Different types of cancellations in a document.

important to identify such sections while processing the documents. Some types of scribbles and cancellations while writing are presented in Figure 1.7.

## 1.2.2 Segmentation

Segmenting a text [16–20] into individual tokens is critical in the case of OCR. Segmenting handwritten texts for different Indic languages is a difficult task [21, 22] because of commonalities among different characters. There are also non-connected components in characters like "ऱ". When observed in isolation, the "." seems to be noise but when seen with the entire character it presents a meaning. Handling these issues is very challenging and although researchers have proposed different techniques for segmentation of characters, lines and words in different Indic languages [23–25], a full-fledged segmentation scheme has yet to be developed. Segmentation is also difficult due to intra-writer variation as discussed in the previous section. The same is illustrated in Figure 1.8.

## 1.2.3 Applications

Optical character recognition is a vast field with a huge number of applications. One can find OCR systems using state-of-the-art approaches to different languages such as Bangla, English, Latin, Cyrillic, Arabic, Hebrew, Indic, Bengali (Bangla), Devanagari, Tamil, Chinese, Japanese, Korean and many more. Some widely used applications can be seen in the following fields:

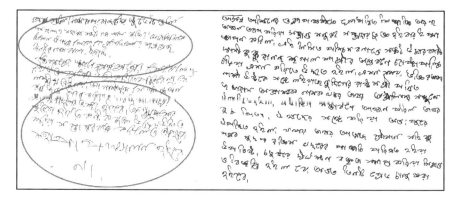

**FIGURE 1.8**
Disparities in handwriting in a document.

legal, banking, healthcare and business. OCR can also be used in many other fields like institutional repositories and digital libraries, CAPTCHA, automatic plate number recognition, optical music recognition without any human effort, handwriting recognition and others [26]. Some of these applications are discussed in the following sections.

### 1.2.3.1 Legal Industry

In the legal industry, OCR plays a significant role in digitizing paper documents. To save space, documents are scanned and entered into computer databases. OCR simplifies the process and makes the documents text-searchable [27, 28].

### 1.2.3.2 Banking

One of the most widely used applications of OCR is in banking. It can process a cheque without human involvement. When a cheque is inserted into a machine, the text written on it is scanned instantly and the correct amount of money is transferred directly to the specified account number. This technique [27, 28], however, is preferable for printed text because it is only partially accurate for handwritten texts.

### 1.2.3.3 Healthcare

OCR also contributes to healthcare by reducing paperwork by a huge volume. Healthcare professionals generally deal with a large number of forms with patient data. This task becomes easier when relevant data are put into an electronic database, where they can be accessed whenever needed. Here, OCR is used to extract information from the forms and put them into the databases [27, 28].

### *1.2.3.4 CAPTCHA*

Nowadays, CAPTCHA is a very common term. Many human activities like transactions, admissions and registrations are carried out on the Internet and need a password to prevent misuse by hackers [29]. With CAPTCHA, an image consisting of a series of letters and numbers is generated which is obscured by image distortion techniques, size and font variation, distracting backgrounds, random segments, highlights and noise in the image. OCR is useful here in reducing the noise and segmenting the image to make it traceable [26].

### *1.2.3.5 Automatic Number Recognition*

Automatic plate number recognition systems use OCR to identify vehicle registration plates.

### *1.2.3.6 Handwriting Recognition*

Handwriting recognition systems or HOCR [30–32] are able to receive and interpret handwritten input from different sources such as paper documents, images, etc. The image may be sensed "offline" or "online" [26].

## 1.3 Natural Language Processing

Natural language processing (NLP) [33] involves the understanding of what is written in a text. Once the characters are identified in a text, NLP-based approaches then come into play to actually understand the meaning of what is written. NLP has several applications including text summarization [34], question answering [35], text categorization [36, 37], sentiment analysis [38] and word sense disambiguation [39]. Research in different non-Indic languages has progressed significantly [40, 41], but for Indic languages it has not. One reason for this is the unavailability of datasets. The Indic languages are also very complex both from grammatical and linguistic points of view [42]. Handling Indic texts is also a problem as extended font sets in UTF-8 are required, but not always supported by different platforms.

NLP has several important stages which are detailed in the following subsections.

### 1.3.1 Tokenization

Tokenization is the task of segmenting a string into corresponding words. It is one of the first steps in different NLP-based applications and can be done at the sentence, word or character level. The different schemes are enumerated in Table 1.2 for the sentence, "Today is Friday. Tomorrow is Saturday". The tokenization is performed by discarding spaces. The same can be done by involving spaces as well.

**TABLE 1.2**

Levels of Tokenization

| Tokenization Level | Output |
| --- | --- |
| Word level | "Today", "is", "Friday", "Tomorrow", "is", "Saturday" |
| Sentence level | "Today is Friday", "Tomorrow is Saturday" |
| Character level | "T", "o", "d", "a", "y", "i", "s" and so on |

## 1.3.2 Stop Word Removal

Irrelevant words, characters and punctuation marks within a text are termed stop words. Stop words are very much context-dependent, i.e. a certain set of words or punctuations may not be important for a particular type of application but of extreme importance for a different domain. For example, in the case of text summarization, the punctuation marks are not very important and thus may be discarded. However, if the problem involves sentiment analysis, the punctuation is very important. Suppose there are two sentences: "You will go" and "You will go?". Even though the two sentences have the same exact words, one is a declarative statement while the other is an interrogative statement.

## 1.3.3 Stemming

Stemming involves the derivation of a root word from the used form of a word. This is presented below.

Root form: "খেলা"

Used forms: "খেলছি", "খেলছিলাম", "খেলতে", etc.

Such different forms of root words are used interchangeably in sentences like:

আমি খেলছি

কাল ফুটবল খেলছিলাম

আজ খেলতে যাবো

If all such forms of the words are used, the number of combinations will be huge. In order to deal with this, stemming may be a solution in which the root form of a word is obtained from the used forms.

## 1.3.4 Part of Speech Tagging

Part of speech tagging is the annotation of each word in the sentence with its part of speech. Often, the part of speech portrays a lot of useful information which can be used for analysis. Two different parts of speech for the same word are presented as follows:

"আকাশের রঙ নীল"

"আমার বন্ধুর নাম নীল"

Here, "নীল" is used as an adjective in the first sentence but as a noun in the second.

### 1.3.5 Parsing

Parsing is the analysis of a sentence with respect to the syntactic structure of the language. It is very helpful in determining the part of speech of words as well as in different applications like word sense disambiguation.

### 1.3.6 Applications

Some of the disparate applications of natural language processing are detailed hereafter.

#### 1.3.6.1 Text Summarization

Text summarization [43, 44] is the technique of extracting the most relevant information from a given text. The term "relevant text" is problem-specific. A text may contain information related to various topics. Summarization generally involves distillation of a particular type of information from a text. It is important to discard the irrelevant information.

#### 1.3.6.2 Question Answering

Question answering [45, 46] involves automated responses to different queries raised by a user. It may be used in healthcare-related applications as well as automated tutoring systems. It is also required in automated customer handling systems. Such a system involves analysis of the texts in order to understand what a user means to say and then generation of the appropriate reply.

#### 1.3.6.3 Text Categorization

Text categorization [47, 48] is the task of grouping texts based on certain criteria. The texts may be grouped based on domains [36] like sports, medical, etc. This is very important as the number of documents on the Internet has increased tremendously. It is very important to categorize and order such information for its efficient retrieval, thus reducing the workload on the system.

#### 1.3.6.4 Sentiment Analysis

Sentiment analysis [49, 50] is the process of analyzing different types of information within a text, like emotional- and opinion-related information. It can help by automatically suggesting the importance of certain texts which at times may be overlooked due to human error.

#### 1.3.6.5 Word Sense Disambiguation

Word sense disambiguation [51, 52] is the process of determining the actual meaning of a word in a sentence when a word has multiple meanings. For instance, the word "bank" may mean "river bank" or "a place where we deposit money". It is very important to understand the meaning of different words prior to understanding the meaning of a text for different applications like summarization and evaluation. For instance, in the sentence, "I will go to the bank for a loan", "bank" refers to the financial place and not to a river bank. The meaning of a word is very context-dependent, and analysis of neighboring text can help derive its meaning.

## 1.4 Conclusion

Artificial intelligence has developed significantly over the years and has provided different simplified solutions. Document image processing has also advanced with the passage of time. Document processing involves both recognition and understanding. Optical character recognition is required for recognition of characters, words and sentences from document images which is followed by natural language processing for the purpose of understanding. Research in these fields for different non-Indic languages has progressed significantly, while for Indic languages it has not. In this chapter, we have presented some avenues for OCR and NLP which are critical in developing a full-fledged system which can truly understand a document in an Indic language.

## References

1. Alzou'bi, S., Alshibl, H., & Al-Ma'aitah, M. (2014). Artificial intelligence in law enforcement, a review. *International Journal of Advanced Information Technology*, 4(4), 1.
2. Dhar, A., Dash, N. S., & Roy, K. (2018). Categorization of Bangla web text documents based on TF-IDF-ICF text analysis scheme. In: *2018 52nd Annual Convention of Computer Society of India (CSI-2017)*, pp. 477–484.
3. Bhunia, A. K., Das, A., Roy, P. P., & Pal, U. (2015, August). A comparative study of features for handwritten Bangla text recognition. In: *2015 13th International Conference on Document Analysis and Recognition (ICDAR)*, pp. 636–640. IEEE.
4. Bag, S., Harit, G., & Bhowmick, P. (2014). Recognition of Bangla compound characters using structural decomposition. *Pattern Recognition*, 47(3), 1187–1201
5. Rakshit, P., Halder, C., Ghosh, S., & Roy, K. (2018). Line, word, and character segmentation from Bangla handwritten text—A precursor toward Bangla HOCR. In: *Advanced Computing and Systems for Security* (pp. 109-120). Springer, Singapore.
6. Manning, C. D., Manning, C. D., & Schütze, H. (1999). *Foundations of Statistical Natural Language Processing*. MIT Press, Cambridge, MA.

7. Jurafsky, D. (2000). *Speech and Language Processing: An Introduction to Natural Language Processing. Computational Linguistics, and Speech Recognition.* Prentice Hall, Upper Saddle River, NJ

8. Collobert, R., Weston, J., Bottou, L., Karlen, M., Kavukcuoglu, K., & Kuksa, P. (2011). Natural language processing (almost) from scratch. *Journal of Machine Learning Research*, 12(Aug), 2493–2537.

9. http://www.cvisiontech.com/products/maestro-recognition-server.html. Visited on 1.11.2018.

10. https://acrobat.adobe.com/in/en/acrobat/how-to/convert-jpeg-tiff-scan-to-pdf.html. Visited on 1.11.2018

11. https://www.clarabridge.com/nlp-natural-language-processing/. Visited on 1.11.2018.

12. Obaidullah, S. M., Halder, C., Santosh, K. C., Das, N., & Roy, K. (2018). PHDIndic_11: page-level handwritten document image dataset of 11 official Indic scripts for script identification. *Multimedia Tools and Applications*, 77(2), 1643–1678.

13. Santosh, K. C., & Wendling, L. (2015). Character recognition based on non-linear multi-projection profiles measure. *Frontiers of Computer Science*, 9(5), 678–690.

14. Santosh, K. C. (2011, September). Character recognition based on dtw-radon. In: *2011 International Conference on Document Analysis and Recognition* (pp. 264–268). IEEE.

15. Obaidullah, S. M., Santosh, K. C., Halder, C., Das, N., & Roy, K. (2017). Automatic Indic script identification from handwritten documents: Page, block, line and word-level approach. *International Journal of Machine Learning and Cybernetics*, 10(1), 1–20.

16. Hangarge, M., Santosh, K. C., & Pardeshi, R. (2013, August). Directional discrete cosine transform for handwritten script identification. In: *2013 12th International Conference on Document Analysis and Recognition* (pp. 344–348). IEEE.

17. Obaidullah, S. M., Goswami, C., Santosh, K. C., Das, N., Halder, C., & Roy, K. (2017). Separating Indic scripts with matra for effective handwritten script identification in multi-script documents. *International Journal of Pattern Recognition and Artificial Intelligence*, 31(05), 1753003.

18. Obaidullah, S. K., Santosh, K. C., Halder, C., Das, N., & Roy, K. (2017). Word-level multi-script Indic document image dataset and baseline results on script identification. *International Journal of Computer Vision and Image Processing (IJCVIP)*, 7(2), 81–94.

19. Obaidullah, S. M., Bose, A., Mukherjee, H., Santosh, K. C., Das, N., & Roy, K. (2018). Extreme learning machine for handwritten Indic script identification in multiscript documents. *Journal of Electronic Imaging*, 27(5), 051214.

20. Obaidullah, S. M., Santosh, K. C., Das, N., Halder, C., & Roy, K. (2018). Handwritten Indic script identification in multi-script document images: A survey. *International Journal of Pattern Recognition and Artificial Intelligence*, 32(10), 1856012.

21. Halder, C., & Roy, K. (2011, December). Word & character segmentation for Bangla handwriting analysis & recognition. In: *2011 Third National Conference on Computer Vision, Pattern Recognition, Image Processing and Graphics (NCVPRIPG)* (pp. 243–246). IEEE.

22. Obaidullah, S. M., Das, N., Halder, C., & Roy, K. (2015, July). Indic script identification from handwritten document images—An unconstrained block-level approach. In: *2015 IEEE 2nd International Conference on Recent Trends in Information Systems (ReTIS)* (pp. 213–218). IEEE.

23. Louloudis, G., Gatos, B., Pratikakis, I., & Halatsis, C. (2009). Text line and word segmentation of handwritten documents. *Pattern Recognition, 42*(12), 3169–3183.

24. Nicolaou, A., & Gatos, B. (2009, July). Handwritten text line segmentation by shredding text into its lines. In: *10th International Conference on Document Analysis and Recognition, 2009. ICDAR'09* (pp. 626–630). IEEE.

25. Bishnu, A., & Chaudhuri, B. B. (1999, September). Segmentation of Bangla hand-written text into characters by recursive contour following. In: *Proceedings of the Fifth International Conference on Document Analysis and Recognition, 1999. ICDAR'99* (pp. 402–405). IEEE.

26. Arif Mir Asif Ali Mir, Abdul Hannan Shaikh, Perwej Yusuf and Arjun Vithalrao Mane. (2014) An overview and applications of optical character recognition. *International Journal of Advance Research in Science and Engineering, 3*(7).

27. http://www.cvisiontech.com/reference/general-information/ocr-applications. html. Visited on 10.10.2018.

28. Ganis, M. D., Wilson, C. L., & Blue, J. L. (1998). Neural network-based systems for handprint OCR applications. *IEEE Transactions on Image Processing, 7*(8), 1097–1112.

29. Gossweiler, R., Kamvar, M., & Baluja, S. (2009). *What's Up CAPTCHA? A CAPTCHA Based on Image Orientation*. WWW.

30. Santosh, K. C., Nattee, C., & Lamiroy, B. (2010, November). Spatial similarity based stroke number and order free clustering. In: *2010 12th International Conference on Frontiers in Handwriting Recognition* (pp. 652–657). IEEE.

31. Santosh, K. C., Nattee, C., & Lamiroy, B. (2012). Relative positioning of stroke-based clustering: A new approach to online handwritten devanagari character recognition. *International Journal of Image and Graphics, 12*(02), 1250016.

32. Santosh, K. C., & Iwata, E. (2013). *Stroke-Based Cursive Character Recognition*. arXiv preprint arXiv:1304.0421.

33. Sager, N., Siegel, J. S., & Larmon, E. A. (1981) *Natural Language Information Processing: A Computer Grammar of English and Its Applications*. Reading, MA: Addison-Wesley.

34. Saggion, H., Bontcheva, K., & Cunningham, H. (2003, April). Robust generic and query-based summarisation. In: *Proceedings of the Tenth Conference on European Chapter of the Association for Computational Linguistics* (Volume 2, pp. 235–238). Association for Computational Linguistics.

35. Ravichandran, D., & Hovy, E. (2002, July). Learning surface text patterns for a question answering system. In: *Proceedings of the 40th Annual Meeting on Association for Computational Linguistics* (pp. 41–47). Association for Computational Linguistics.

36. Dhar, A., Dash, N. S., & Roy, K. (2017). Application of TF-IDF feature for catego-rizing documents of online Bangla web text corpus. In: *Proceedings of FICTA* (pp. 51–59).

37. Dhar, A., Dash, N. S., & Roy, K. (2018). Categorization of Bangla web text doc-uments based on TF-IDF-ICF text analysis scheme. In: *Proceedings of CSI* (pp. 477–484).

38. Kouloumpis, E., Wilson, T., & Moore, J. D. (2011). Twitter sentiment analysis: The good the bad and the omg! *Icwsm, 11*(538–541), 164.

39. Yarowsky, D. (1995, June). Unsupervised word sense disambiguation rivaling supervised methods. In: *Proceedings of the 33rd Annual Meeting on Association for Computational Linguistics* (pp. 189–196). Association for Computational Linguistics.

40. Bijalwan, V., Kumar, V., Kumari, P., & Pascual, J. (2014). KNN based machine learning approach for text and document mining. *International Journal of Database Theory and Application, 7*(1), 61–70.
41. Wu, K., Zhou, M., Lu, X. S., & Huang, L. (2017, October). A fuzzy logic-based text classification method for social media data. In: *2017 IEEE International Conference on Systems, Man, and Cybernetics (SMC)* (pp. 1942–1947). IEEE.
42. Dhar, A., Dash, N. S., & Roy, K. (2018). A fuzzy logic-based Bangla text classification for web text documents. *Journal of Advanced Linguistic Studies, 7*, 159–187.
43. Lloret, E., & Palomar, M. (2012). Text summarisation in progress: A literature review. *Artificial Intelligence Review, 37*(1), 1–41.
44. Benbrahim, M., & Ahmad, K. (1995) Text summarisation: The role of lexical cohesion analysis. *The New Review of Document & Text Management, 1*, 321–335.
45. Lin, D., & Pantel, P. (2001). Discovery of inference rules for question-answering. *Natural Language Engineering, 7*(4), 343–360.
46. Brill, E., Dumais, S., & Banko, M. (2002, July). An analysis of the AskMSR question-answering system. In: *Proceedings of the ACL-02 Conference on Empirical Methods in Natural Language Processing* (Volume 10, pp. 257–264). Association for Computational Linguistics.
47. Joachims, T. (1998, April). Text categorization with support vector machines: Learning with many relevant features. In: *European Conference on Machine Learning* (pp. 137–142). Springer, Berlin, Heidelberg.
48. Yang, Y., & Liu, X. (1999, August). A re-examination of text categorization methods. In: *Proceedings of the 22nd Annual International ACM SIGIR Conference on Research and Development in Information Retrieval* (pp. 42–49). ACM.
49. Li, N., & Wu, D. D. (2010). Using text mining and sentiment analysis for online forums hotspot detection and forecast. *Decision Support Systems, 48*(2), 354–368.
50. Gilbert, C. H. E. (2014). Vader: A parsimonious rule-based model for sentiment analysis of social media text. In: *Eighth International Conference on Weblogs and Social Media (ICWSM-14).* http://comp. social. gatech. edu/papers/icwsm14. vader. hutto. pdf. Available at 20/04/16.
51. Yarowsky, D. (1995, June). Unsupervised word sense disambiguation rivaling supervised methods. In: *Proceedings of the 33rd annual meeting on Association for Computational Linguistics* (pp. 189–196). Association for Computational Linguistics.
52. Banerjee, S., & Pedersen, T. (2002, February). An adapted Lesk algorithm for word sense disambiguation using WordNet. In: *International Conference on Intelligent Text Processing and Computational Linguistics* (pp. 136–145). Springer, Berlin, Heidelberg.

# 2

# An Approach toward Character Recognition of Bangla Handwritten Isolated Characters

Payel Rakshit, Chayan Halder and Kaushik Roy

## CONTENTS

## 2.1 Introduction

Recognition of characters from their images is one of the most essential tasks in the field of computer vision. Optical character recognition (OCR) has lately become a very interesting field of research, especially for Indic scripts, due to its potential in Asian countries like India. The task, however, is quite difficult when images of handwritten characters are considered [1, 2]. Recently, various works on character recognition from handwritten characters have become available [1–13], while some very established standard character recognition systems from printed characters are already commercially available [14, 15]. Numerous techniques on several scripts [3] for offline isolated character recognition have already appeared in the literature. Bhowmik et al. [3] proposed two-stage hierarchical learning architectures based on a support vector machine (SVM), whereas Bhattacharya et al. [5] proposed an efficient two-stage approach for Bangla handwritten character recognition and achieved very good recognition accuracy with a multi-layer perceptron (MLP) classifier. Maitra et al. [6] also developed an approach for the handwritten character recognition of multiple scripts based on a convolution neural network (CNN). At the same time, Surinta et al. [7] contributed a system that used local gradient feature descriptors. Another method using soft computing paradigm embedded with a two-pass approach was developed by Das et al. [8]. Sarkhel et al. [9] proposed a multi-objective approach for isolated Bangla character and digit recognition. Pramanik and Bag [10] came up with an idea for Bangla compound character recognition based on shape decomposition. Sarkhel et al. [11] also presented a technique for the recognition of popular Indic scripts. Recently, a comparative performance study was made on feature selection techniques for offline OCR by Kumar et al [12]. Additionally, isolated handwritten character recognition from different scripts such as Devanagari, Bangla, Oriya and Japanese-Katakana were attempted by Santosh and Wendling [13]. The Radon transform was used to produce a multi-projection profile of characters and dynamic time warping (DTW) was used to match that profile.

The available approaches towards handwritten character recognition usually consist of several steps. Among them, two major steps are feature extraction and classifier design. From the related survey, some well-known and popularly used classifiers can be found, including multilayer perceptron (MLP) [5], radial basis function (RBF) [3], modified quadratic discriminant function (MQDF) [5], support vector machine (SVM) [3], nearest neighbour [3], etc. Some frequently used and well-known features of state-of-the-art approaches are chain code direction, gradient, curvature, the Gabor transform, wavelet transform and statistical/structural features, etc. [6]. Recently, in a few character recognition/document recognition studies, CNNs [6] were used to achieve higher accuracy. The literature reveals that although there are methods available for character recognition, there is still room for excellence in recognition performance. In this chapter, an approach to recognizing handwritten, isolated Bangla characters has been studied on a relatively large database. Along with this, how to improve the performance of existing methods is described, using some standard databases of isolated Bangla handwritten characters. There are two objectives of these tasks which need simultaneous optimization: maximizing the recognition accuracy and minimizing the effective cost.

Section 2.2 presents a proposed framework where database, feature and classification approaches are described; results and a discussion, using a comparative study with state-of-the-art methods, are presented in Section 2.3; and, lastly, Section 2.4 contains our conclusions.

## 2.2 Proposed Framework

An optical character recognition system usually has various steps to follow before it can actually recognize characters. The preliminary steps include cleaning and normalizing the raw input data, for which some pre-processing steps need to be performed. In this proposed method of character recognition

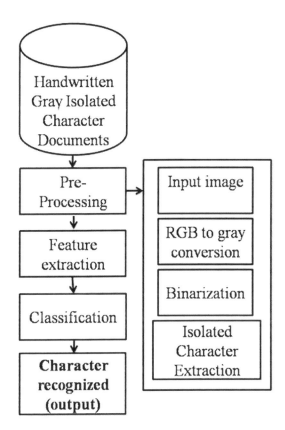

**FIGURE 2.1**
Block diagram of the proposed character recognition system of isolated characters.

from handwritten Bangla isolated characters, the raw collected handwritten characters need initially to be cleaned and extracted from the images of data collection forms to create a database of isolated character images. This pre-processing step is similar in our previous work [16]. After that, feature extraction and classification methods are followed for recognition of the isolated characters. The block diagram of Figure 2.1 will provide a better understanding of the procedure. The following sections provide all the details of the steps that are performed for recognition.

## 2.2.1 Database

In this proposed work, a previously collected database of handwritten, isolated Bangla characters of 250 writers is used, consisting of 89,815 isolated Bangla character images out of which 64,515 are letters, 12,650 are numerals and 12,650 are vowel modifiers. The other standard databases that are used for comparative study are the ISI handwritten character database [17] and CMATERdb 3.1.2 [8]. Details of the databases are provided in Table 2.1.

**TABLE 2.1**

Details of the Databases Used in the Proposed Work

| Database | Total Writers | Total Characters |
|---|---|---|
| Our own database | 250 | 89,815 |
| ISI handwritten character database [17] | 20 | 1,500 |
| CMATERdb 3.1.2 [8] | 240 | 12,000 |

## 2.2.2 Feature Extraction

The feature extraction stage of this proposed work is comprised of the 64-dimensional contour chain code-based feature, which is extracted from isolated gray images of characters/numerals. The computational technique of 64-dimensional feature is provided in the following.

Firstly, binarization of the input gray image is done and the bounding boxes are computed. After this, the contour points of each bounding box are considered. Then, $7 \times 7$ blocks are created out of the bounding box by dividing them equally. The four directional chain codes are computed for each contour point in each of these blocks along with their frequency. In this way, for each block, four integer values are computed representing the frequencies of chain code of four directions which produces a $7 \times 7 \times 4 = 196$ dimensional feature. Then, the Gaussian filter is applied on each of the blocks to down-sample these blocks to 4x4 blocks, which finally produces the $4 \times 4 \times 4 = 64$-dimensional chain code feature.

## 2.2.3 Attribute Selection and Classification

*Attribute selection*: Attribute selection is important for improving the performance of a system in terms of computational complexity. Also, it reduces some redundant attributes which may reduce accuracy. Here, in this work, the filter type feature selection method is used as it will not modify the given feature set to some reduced feature set that is specifically tuned for any particular classifier. The feature set reduction is used for only CMATERdb 3.1.2. A filtered correlation-based feature subset selection method is applied with a greedy stepwise search to reduce dimensionality. The selection method is more effective than wrapper methods due to its less computational overhead and generalized selection of features that are already available. In this work, the feature reduction has reduced the feature dimension from 64 to 47. The details of attribute selection and filtered correlation-based feature subset selection method can be found in [18].

*Classification*: Three different classification approaches are applied for these three databases. Random forest, simple logistic and instance-based

K-NN (IBK) classification methods are used to get the best recognition performances of all the available databases.

- **Random forest:** This is one of several supervised learning algorithms that creates different numbers of decision trees (thus creating a forest of decision trees). It chooses different feature attributes randomly to create different decision trees during training time. Rather than searching for the most important feature, it searches for the best feature among the pool using a bagging method which results in a better model for classification. This makes the classifier more stable and accurate compared to other contemporary classifiers. In the testing phase, during predictions of unseen samples, an average value is taken from the predictions of all the individual regression trees or majority vote is used for final decision-making. More details can be found in [19].

- **Instance-based K-NN (IBK):** The usual K-NN classification technique looks for the K nearest neighbours to which the new instance of the current class can be assigned. When the value of K is 1, it always checks the immediate nearest neighbour and is incremental in nature. Usually, it does not rely on the training phase or, in other words, it does not have so much to do during training; an efficient training phase will do some indexing to reduce the time complexity during testing phase where exhaustive comparisons are required to find the nearest neighbour of a new instance of an unknown class. The instance-based K-NN (IBK) classification is an extension of the usual K-NN, where specific instances are used to improve the performance of the supervised K-NN. This type of technique does not maintain a set of abstractions derived from specific instances. It reduces the problem of the large storage requirement of the conventional K-NN approach. The IBK is a supervised learning algorithm which learns from labelled examples of instances. The details can be found in [20].

- **Simple logistic:** The LogitBoost algorithm and simple regression functions are the base learning mechanism for fitting the logistic model into a simple logistic classification. The dimension of the input feature vector set and its mapping to the output are same as that of the MLP classifier. It uses cross-validation to determine the optimal number of LogitBoost iterations that is primarily used for the automatic selection of attributes. More details can be found in [18].

- **Multi-layer perceptron (MLP):** This is the most widely used classifier for this type of research, consisting mainly of three layers of feedforward artificial neurons represented as nodes. Each node of the MLP is connected to another by a single connector whose

strength or capacity is represented by the level or weight associated with it. Generally the weight vector and the bias vector of
each layer pair are trained by a back propagation algorithm. The
input layer neurons are determined by the feature dimension,
and the number of classes it needs to be mapped determines the
output layer neuron count. Neurons in the hidden layers (H) are
determined during training phase. In our case, it is calculated as
H=total attributes + total classes. The momentum and learning rate
are chosen to be 0.7468 and 0.8907, respectively. Further details can
be found in [19].

## 2.3  Results and Discussion

This section describes the experimental setup and the outcomes of the proposed method. As described earlier, in this proposed work our own collected
database of more than 89,000 Bangla characters is used. In our experimental
scenario, the k-fold cross validation method is used, where K is chosen to be 5,
factually. The contour chain code-based feature with a random forest classifier
has yielded recognition accuracy of 99.64% for these isolated characters (out of
89,815 characters, 323 characters were found to be incorrectly classified). The
50 basic characters (letters) are considered along with all the numerals and ten
vowel modifiers of the Bangla script in this proposed work. In classification,
the true positive ($T_p$), false positive ($F_p$), true negative ($T_n$), false negative ($F_n$) are
the measures of result outcomes or, in other words, observations. True positive
means the classification result is correctly indicating a positive class instance
as a positive one. True negative means the classification result is correctly indicating a negative class instance as a negative one. A false positive means the
classification result is incorrectly indicating a negative class instance as a positive one. A false negative means the classification result is incorrectly indicating a positive class instance as a negative one. Precision (P), recall (R) and $F_1$
score ($F_1$) are the most widely used statistical measures for judging recognition
performances. Precision is the measure of how many relevant instances are
retrieved. This is also known as the positive predictive value.

$$\text{Precision (P)} = \frac{|(\text{Relevent instances}) \cap (\text{Retrieved instances})|}{|(\text{Retrieved instances})|}$$

This can also be defined as:

$$P = \frac{T_p}{T_p + F_p}$$

Recall is a measure of how many truly relevant instances are retrieved among all the relevant instances.

$$\text{Recall (R)} = \frac{|(\text{Relevent instances}) \cap (\text{Retrieved instances})|}{|(\text{Relevent instances})|}$$

This can also be defined as:

$$R = \frac{T_p}{T_p + F_n}$$

$F_1$ score $(F_1)$ is the measure of accuracy of the observations which can be defined as the harmonic mean of precision and recall scores.

$$F_1 = 2\frac{P \times R}{P + R}$$

Table 2.2 provides these scores for each individual character of Bangla. The table shows that character এ has the best precision recall score of 1 which in turn produced an $F_1$ score of 1. The lowest precision score of 0.979 is given by the numeral ৬, while the vowel modifier ী provides the lowest recall score of 0.987. For both classes the $F_1$ score is also low, while the numeral ৬ is the lowest (0.988) and the vowel modifier ী and numeral ৭ the second lowest (0.991) amongst all the characters. The reduction in precision and recall can be due to the fact that the numeral ৬ and the letter ড are very close in terms of the writing pattern and the same can be observed for the vowel modifier ী and numeral ৭, as there can be confusion because of the structural pattern of these two, which can be observed in Figure 2.2. The average precision, recall and $F_1$ scores of all the characters are 0.996385714 ($\approx$ 0.9964), 0.996385714 ($\approx$ 0.9964) and 0.996414286 ($\approx$ 0.9964), respectively.

## 2.3.1 Comparative Study

This section deals with the comparative study of our method with state-of-the-art methods that have already been reported in literature. The methods reported by Bhattacharya et al. [5] and Das et al. [21] are also on isolated basic character recognition. During comparison, two types of comparative analysis can be done: (1) a comparative study of the reported results regardless of the database and methods used, which can be termed indirect comparison; and (2) considering one of the aspects to be constant, like database or method, during the comparison, which can be termed a direct comparison of the works.

1. *Indirect comparison*: A comparative study with other similar works on Bangla isolated basic characters shows that our method has outperformed other methods by a large margin in terms of accuracy, with

**TABLE 2.2**

Precision, Recall and F-Measure Scores of
Individual Bangla Characters

| Characters | Precision | Recall | $F_1$ score |
| --- | --- | --- | --- |
| 0 | 0.995 | 0.998 | 0.996 |
| ১ | 0.992 | 0.999 | 0.995 |
| ২ | 0.994 | 0.999 | 0.997 |
| ৩ | 0.979 | 0.997 | 0.988 |
| ৪ | 0.995 | 1 | 0.998 |
| ৫ | 0.991 | 0.998 | 0.995 |
| ৬ | 0.991 | 0.995 | 0.993 |
| ৭ | 0.989 | 0.994 | 0.991 |
| ৮ | 0.997 | 0.999 | 0.998 |
| ৯ | 0.998 | 0.996 | 0.997 |
| অ | 0.995 | 0.997 | 0.996 |
| আ | 0.993 | 0.997 | 0.995 |
| ই | 0.995 | 0.997 | 0.996 |
| ঈ | 0.996 | 0.996 | 0.996 |
| উ | 0.997 | 0.995 | 0.996 |
| ঊ | 0.998 | 0.997 | 0.998 |
| ঋ | 0.999 | 0.998 | 0.998 |
| এ | 0.998 | 0.997 | 0.997 |
| ঐ | 0.995 | 0.997 | 0.996 |
| ও | 0.998 | 0.995 | 0.997 |
| ঔ | 0.996 | 0.995 | 0.996 |
| ক | 0.995 | 0.999 | 0.997 |
| খ | 0.998 | 0.999 | 0.998 |
| গ | 0.995 | 0.996 | 0.996 |
| ঘ | 0.995 | 0.996 | 0.996 |
| ঙ | 0.997 | 0.997 | 0.997 |
| চ | 0.998 | 0.998 | 0.998 |
| ছ | 0.997 | 0.999 | 0.998 |
| জ | 0.998 | 0.998 | 0.998 |
| ঝ | 0.997 | 0.997 | 0.997 |
| ঞ | 0.998 | 0.998 | 0.998 |
| ট | 0.997 | 0.995 | 0.996 |
| ঠ | 1 | 0.997 | 0.998 |
| ড | 0.999 | 0.992 | 0.996 |
| ঢ | 0.997 | 0.998 | 0.998 |
| ণ | 0.996 | 0.997 | 0.996 |
| ত | 0.996 | 0.991 | 0.993 |
| থ | 0.997 | 0.998 | 0.998 |
| দ | 0.997 | 0.997 | 0.997 |

*(Continued)*

**TABLE 2.2 (CONTINUED)**

Precision, Recall and F-Measure Scores
of Individual Bangla Characters

| Characters | Precision | Recall | $F_1$ score |
|---|---|---|---|
| শ | 0.000 | 0.998 | 0.998 |
| ন | 0.996 | 0.998 | 0.997 |
| প | 0.995 | 0.996 | 0.995 |
| ফ | 0.995 | 0.997 | 0.996 |
| ব | 0.998 | 0.998 | 0.998 |
| ভ | 0.998 | 0.991 | 0.995 |
| ম | 0.996 | 0.997 | 0.997 |
| য | 0.996 | 0.994 | 0.995 |
| র | 0.998 | 0.993 | 0.996 |
| ল | 1 | 1 | 1 |
| শ | 0.998 | 0.993 | 0.996 |
| ষ | 0.998 | 0.998 | 0.998 |
| স | 1 | 0.995 | 0.998 |
| হ | 1 | 0.997 | 0.998 |
| ড় | 0.999 | 0.998 | 0.998 |
| ঢ় | 0.996 | 0.996 | 0.996 |
| য় | 0.997 | 0.992 | 0.995 |
| ৎ | 0.998 | 1 | 0.999 |
| ং | 0.999 | 0.997 | 0.998 |
| ঃ | 0.998 | 0.998 | 0.998 |
| ঁ | 0.998 | 0.995 | 0.997 |
| া | 0.993 | 0.992 | 0.993 |
| ি | 0.996 | 0.998 | 0.997 |
| ী | 0.995 | 0.987 | 0.991 |
| ু | 0.998 | 0.994 | 0.996 |
| ৃ | 0.996 | 0.994 | 0.995 |
| ় | 0.996 | 0.997 | 0.996 |
| ে | 0.998 | 0.995 | 0.996 |
| ৈ | 0.997 | 0.995 | 0.996 |
| ো | 0.999 | 0.998 | 0.999 |
| ্ | 0.999 | 0.998 | 0.998 |

**FIGURE 2.2**
Some handwritten letters, numerals and vowel modifiers of Bangla with similar shape patterns.

a bigger database size and higher number of classes (numerals and vowel modifiers are included along with the basic characters) than in the previous state-of-the-art methods. The methods addressed by the comparative study are those of Bhattacharya et al. [5] and Das et al. [21]. Two standard databases of basic isolated characters, namely the ISI handwritten character database [17] and CMATERdb 3.1.2 [8], are used by [5] and [21] respectively. The details of the databases are provided in Section 2.2. Table 2.3 shows the comparative result analysis of the different methods found in literature. From the table, it is clear that our method has fewer feature dimensions and a single classification approach, and still achieves better performance in terms of recognition. If the training time is considered for the determination of computational complexity, it will be lower than the rest, as it has fewer feature dimensions. One can argue that no direct comparison is made, here, with the other works in terms of database and computational cost; to counter that, a direct comparison to determine the robustness of the method follows.

2. *Direct comparison*: For direct comparison, we have applied our feature on the ISI handwritten character database [17] and CMATERdb 3.1.2 [8]. After that, the best classification outcomes are studied for those databases for which feature reduction and different classification methods are applied. When our feature is applied in the ISI handwritten character database, considering 60 classes (letters + numerals), it is found that this database yields a 98.13% recognition performance. When the same feature is applied in the CMATERdb 3.1.2 database, considering the same 50 classes, a 99.79% recognition performance is the result. The comparative study can be found in Table 2.4. Another view of the performance score can give us a different kind of comparative outcome. If the performances of the methods on these standard databases are compared with the result of previously reported state-of-the-art methods, it is found from Tables 2.5 and 2.6 that, compared to both of the standard databases, our applied method has produced better recognition accuracy. Table 2.5 shows the comparative results of our method with the

**TABLE 2.3**

Comparative Result Analysis of Isolated Character Recognition Methods

| Approaches | Number of Classes | Size of the Database | Feature Dimension | Classification | Recognition Accuracy (%) |
|---|---|---|---|---|---|
| Bhattacharya et al. [5] | 50 | 12,858 | 160 | MQDF + MLP | 95.84 |
| Das et al. [21] | 50 | 12,000 | 132 | MLP | 85.40 |
| Proposed | 70 | 89,815 | 64 | Random Forest | 99.64 |

previously reported results of Bhattacharya et al. [5]: our method is 2.29% better than their reported performance in the ISI handwritten character database [17]. Similarly, from Table 2.6, it is observed that our method has produced a 14.39% better recognition performance than Das et al. [21] reported. In another comparative study of recognition performances considering only the numerals of the ISI handwritten character database [17], again our proposed method outperforms the recognition performance reported by Das et al. [8] by 1.37%. Table 2.7 provides the detail results. It worth mentioning

**TABLE 2.4**

Comparative Study of Current Performances of the Databases

| Database | Number of Classes | Total Instances | Recognition Accuracy (%) |
|---|---|---|---|
| ISI handwritten character database [14] | 60 | 1,500 | 98.13 |
| CMATERdb 3.1.2 [8] | 50 | 12,000 | 99.79 |
| Proposed | 70 | 89,815 | 99.64 |

**TABLE 2.5**

Comparative Study of Current Method and Previously Reported Method of Bhattacharya et al. [5] on ISI Handwritten Character Database [17]

| Method | Number of Classes | Recognition Accuracy (%) |
|---|---|---|
| Bhattacharya et al. [5] | 50 | 95.84 |
| Proposed | 60 | 98.13 |

**TABLE 2.6**

Comparative Study of Current Method and Previously Reported Method of Das et al. [21] on CMATERdb 3.1.2 [8]

| Method | Number of Classes | Recognition Accuracy (%) |
|---|---|---|
| Das et al. [21] | 50 | 85.40 |
| Proposed | 50 | 99.79 |

**TABLE 2.7**

Comparative Study of Current Method and Previously Reported Method of Das et al. [8] on Only Numerals in ISI Handwritten Character Database [17]

| Method | Number of Classes | Recognition Accuracy (%) |
|---|---|---|
| Das et al. [8] | 10 | 98.63 |
| Proposed | 10 | 100 |

that, though the feature is different, we used the same MLP classifier as Das et al. did.

---

## 2.4 Conclusion

This chapter presents a study of character recognition from handwritten isolated characters in different standard databases. Our own database of 89,815 characters (letters + numerals + vowel modifiers) is considered. The standard database of ISI handwritten characters and CMATERdb 3.1.2 are also considered for a comparative study of our methods. The comparative study shows that our method has outperformed every other method in every aspect. Further study also reveals that our proposed feature is robust and stable across all the standard databases. Though the recognition performance is quite high, there are still classification errors in the results, due to the handwritten shape similarity of some Bangla letters, numerals and vowel modifiers.

In future, more features and feature combinations can be studied to find similar kinds of feature descriptors that will provide very high recognition performances across different databases. Similar studies will be performed on handwritten Bangla text databases to get better performances of recognition not only in terms of accuracy but also in terms of computational complexity. Additionally, extreme learning methods similar to the work of [22] and deep learning-based methods similar to [23] may be employed in future on text databases attempts similar to this one. Graph mining-based specific content selection methods similar to [24] may be implemented in future for fast learning. Along with this, we will try to publish a database similar to the script identification database of [25], and make it freely available to the research community for the study of character recognition.

---

## References

1. P. Rakshit, C. Halder, S. Ghosh, K. Roy , "Line, word, and character segmentation from Bangla handwritten text-A precursor towards Bangla HOCR", In: R. Chaki, A. Cortesi, K. Saeed, N. Chaki (eds), *Advanced Computing and Systems for Security, Advances in Intelligent Systems and Computing*, Vol. 666, Springer, 2018.
2. K. C. Santosh, E. Iwata, "Stroke-based cursive character recognition", *Advances in Character Recognition*, editor: Xiaoqing Ding, ISBN 979-953-307-796-2, 2012.
3. T. K. Bhowmik, P. Ghanty, A. Roy, S. K. Parui, "SVM-based hierarchical architectures for handwritten Bangla character recognition", *International Journal on Document Analysis and Recognition (IJDAR)*, 12(2), Pages 97–108, 2009.

4. U. Bhattacharya, B. B. Chaudhuri, "Handwritten numeral databases of Indian scripts and multistage recognition of mixed numerals", *IEEE Transactions on Pattern Analysis and Machine Intelligence*, 31(3), Pages 444–457, 2009.
5. U. Bhattacharya, M. Shridhar, S. K. Parui, P. K. Sen, B. B. Chaudhuri, "Offline recognition of handwritten Bangla characters: An efficient two-stage approach", *Pattern Analysis and Applications*, 15(4), Pages 445–458, 2012.
6. D. S. Maitra, U. Bhattacharya, S. K. Parui, "CNN based common approach to handwritten character recognition of multiple scripts", *13th International Conference on Document Analysis and Recognition (ICDAR)*, Tunis, Pages 1021–1025. doi: 10.1109/ICDAR.2015.7333916, 2015.
7. Olarik Surinta, Mahir F. Karaaba, Lambert R. B. Schomaker, Marco A. Wiering, "Recognition of handwritten characters using local gradient feature descriptors", *Engineering Applications of Artificial Intelligence*, 45, Pages 405–414, 2015.
8. Nibaran Das, Ram Sarkar, Subhadip Basu, Punam K. Saha, Mahantapas Kundu, Mita Nasipuri, "Handwritten Bangla character recognition using a soft computing paradigm embedded in two pass approach", *Pattern Recognition*, 48(6), Pages 2054–2071, 2015.
9. Ritesh Sarkhel, Nibaran Das, Amit K. Saha, Mita Nasipuri, "A multi-objective approach towards cost effective isolated handwritten Bangla character and digit recognition", *Pattern Recognition*, 58, Pages 172–189, 2016.
10. Rahul Pramanik, Soumen Bag, "Shape decomposition-based handwritten compound character recognition for Bangla OCR", *Journal of Visual Communication and Image Representation*, 50, 2018.
11. Ritesh Sarkhel, Nibaran Das, Aritra Das, Mahantapas Kundu, Mita Nasipuri, "A multi-scale deep quad tree based feature extraction method for the recognition of isolated handwritten characters of popular Indic scripts", *Pattern Recognition*, 71, Pages 78–93, 2017.
12. M. Kumar, M. K. Jindal, R. K. Sharma, S. RaniJindal, "Performance comparison of several feature selection techniques for offline handwritten character recognition", *International conference on research in intelligent and computing in engineering (RICE)*, Pages 1–6, 2018.
13. K. C. Santosh, L. Wendling, "Character recognition based on non-linear multiprojection profiles measure", *Frontiers of Computer Science*, 9(5), Pages 1–13, 2015.
14. https://finereaderonline.com/en-us. Last accessed 11 December 2018.
15. S. Rakshit, S. Basu, "Recognition of handwritten Roman script using Tesseract open source OCR engine", *National Conference on (NAQC)*, Pages 141–145, 2008.
16. C. Halder, K. Roy, "Individuality of isolated Bangla characters", *International Conference on Devices, Circuits and Communications (ICDCCom)*, Pages 1–6, 2014.
17. https://www.isical.ac.in/~ujjwal/download/database.html. last accessed 12.12.2018
18. C. Halder, S. M. Obaidullah, K. C. Santosh, K. Roy, "Content independent writer identification on Bangla script: A document level approach", *International Journal of Artificial Intelligence & Pattern Recognition*, 32(9), 1856011, 2018.
19. S. M. Obaidullah, K. C. Santosh, C. Halder, N Das , K. Roy, "Automatic Indic script identification from handwritten documents: page, block, line and word-level approach", *International Journal of Machine Learning and Cybernetics*, Pages 1–20, 2017.
20. D. Aha, D. Kibler, M. Albert, "Instance-based learning algorithms", *Machine Learning*, 6(1), Pages 37–66, 1991.

21. N. Das, S. Basu, R. Sarkar, M. Kundu, M. Nasipuri, D. K. Basu, "An improved feature descriptor for recognition of handwritten Bangla alphabet", *International Conference on Signal and Image Processing*, pp. 451–454, 2009.

22. S. M. Obaidullah, A. Bose, H. Mukherjee, K. C. Santosh, N. Das, K. Roy, "Extreme learning machine for handwritten Indic script identification in multi-script documents", *Journal of Electronic Imaging*, 27(5), 051214, 2018.

23. S. Ukil, S. Ghosh, S. M. Obaidullah, K. Santosh, K. Roy, N. Das, *"Deep Learning for Word-Level Handwritten Indic Script Identification"*, arXiv Preprint ArXiv:1801.01627, 2018.

24. K. C. Santosh, "g-DICE: graph mining-based document information content exploitation", *International Journal on Document Analysis and Recognition (IJDAR)*, 18(4), Pages 337–355, 2015.

25. S. M. Obaidullah, C. Halder, K. C. Santosh, N. Das, K. Roy, "PHDIndic_11:Page-level handwritten document image dataset of 11 official Indic scripts for script identification", *Multimedia Tools and Applications (MTAP)*, 77(2), Pages 1643–1678, 2017.

# 3

## *Artistic Multi-Character Script Identification*

Mridul Ghosh, Himadri Mukherjee, Sk Md Obaidullah,
KC Santosh, Nibaran Das and Kaushik Roy

### CONTENTS

## 3.1 Introduction

Script identification [1–8], a research topic in image processing and pattern recognition, is a very important phase in the automatic processing of document images in an international scenario. This is because script identification is a requirement for OCR [9] without which an OCR system would not be able to automatically read and convert an image into different scripts that are comfortable for the reader. It is not yet possible to devise a universal or generalized OCR for all kinds of scripts, especially in countries like India, whose Indus script consists of very complex graphic symbols. So, a good technique for script identification can lead us to choose the correct OCR package. In the last several years, many works have been proposed for automatic script identification, but none have been able to fully address the factors involved like multi-script inline level, word level, character level, complex backgrounds, broken lines, noisy data, directional complexity, etc. One feature which has so far been unattainable is artistic multi-character script identification, where different scripted characters vary significantly in color, texture, size, and background. Figure 3.1 depicts the flowchart of our proposed work, which tackles this research topic.

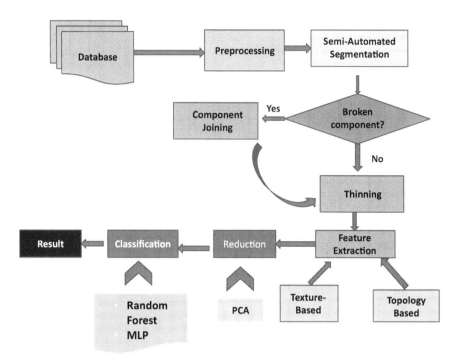

**FIGURE 3.1**
Flowchart of our proposed work.

## 3.2 Literature Review

In [10], authors extracted rotation invariant texture features from an extended version of a multi-channel Gabor filtering method to identify 15 different textures, a technique that has been adopted to identify different scripts, namely Persian, Greek, Russian, Chinese, English, and Malayalam. The authors in [11] proposed a technique for word-wise script identification: their work was carried out in two phases, considering the tradeoff between accuracy and processing speed, in order to identify Bangla, Devanagari, and Roman scripts. In the first phase, a chain-code-histogram was used for feature extraction, returning a 64-dimensional feature at high speed but with less accuracy. In the second phase, considering those samples which yield low accuracy rates, gradient features were extracted from dimension 400. For the classification majority, a voting method was exploited. In [12], Telugu, Devanagari, and English scripts were separated from a printed document using distinct features extracted from the top and bottom profiles of the printed text. The experiment was conducted on 2,400 text lines with a recognition accuracy of 99.67%. The authors of [13] used maximized mutual information to identify the multi-script documents of

Kannada and Hindi, with a dataset of 400 images. They captured the features by edge direction histogram (EDH) and classified them using K nearest neighbor (KNN) and support vector machine (SVM). The authors of [14] used texture feature extraction by Gray-level co-occurrence matrix (GLCM), Gabor energy, wavelet energy, wavelet log mean deviation, wavelet co-occurrence signatures, wavelet log co occurrence signatures, and wavelet scale co-occurrence signatures on 30 different fonts in 100 script images. The authors of [15] discussed the identification of Chinese, Arabic, English, Devanagari, and Bangla scripts: they used features from shape properties which use matras (for Bangla) and shirorekhas (for Devanagari), as well as head lines, base lines, low lines, upper lines, and splits of Bangla and Devanagari lines with zone-wise shape features; for Chinese, English, and Arabic text lines, vertical black run information and Run Length Smoothing Features were considered; and for English and Arabic text line separation, they used a statistical feature based on the distribution of the lowermost points of the components and the water reservoir principal.

## 3.3 Data Collection and Preprocessing

As data collection is the preliminary work in image processing and pattern recognition, research on validating methods is important. However, collecting real-world artistic multi-character scripts to study is not an easy task. From different areas of Kolkata, India, we collected texts not only written in different scripts, but that also varied widely in size, color, texture, background, etc. These scripts alone, however, were not sufficient for our work. So, we made synthetic artistic multi-character scripts/images for experimentation. Until now we have managed to collect only 49 real-world images from various places like company hoardings, banners, t-shirt graffiti, a cinema hall poster, placards, festoons, wall writings, newspapers, etc. We have used mobile phones with resolutions varying from 8 MP to 20 MP, as well as a DSLR still camera (Nikon D3300) with 24.2 MP resolution and an 18–55 mm lens. Around 2628 characters have been extracted from these real-world images, and additionally from 833 synthetic images in which 178 Devanagari, 1599 Bangla, and 851 Roman scripts appear.

Hindi is one of the most widely spoken languages in India with around 200 million people, mainly residing in the northern, western and eastern regions, making use of this language as their communication medium; around 190 million people in West Bengal use this language to communicate too. Though English is a foreign language in India, about 340 million people in our country use it in education institutes, businesses, workplaces, and other sectors.

Figure 3.2 shows various sample images of our dataset. Some of them have a simple background and others a complex background. In (a) the image was

**FIGURE 3.2**
Some sample images from our database. (a) shows a real image, while (b)–(d) are synthetic images.

captured from a banner which is a logo, taken from the roadside where the image was not only written in two different scripts (Devanagari, Roman), but also in an artistic way. In (b), (c), and (d) the images are synthetic artistic images combined with Bangla and Roman: in (b) the image is formed with Bangla, then Roman and Bangla; in (c) Bangla followed by Roman; and in (d) Roman followed by Bangla.

From these images, it can be seen that the images not only vary in color, size, fonts, and textures, but also in their backgrounds (the bag, t-shirt, wall, etc.) In this scenario, detecting a region of interest automatically is a very complicated task. So, to make this step simpler, we detect the region of interest (ROI) manually.

The ROI of each image has been extracted manually, applying Otsu's [16] binarization algorithm to convert the image into a binary, and then a gray, image. The automated segmentation algorithm [17], which uses connected component labeling and disjoint set analysis, has been used to extract individual characters from the bi-script words. In this work, we have considered three different scripts, namely Roman, Devanagari, and Bangla. The words are formed with bi-script characters having $3_{C_2}$ combinations, i.e. Bangla with Devanagari, Roman with Bangla, Devanagari with Roman.

From Figure 3.3, it can be seen that after automatic segmentation only some images are well-segmented, while many are not well-separated. So, to solve this problem, we went for manual segmentation of those images that were left un-segmented in automatic segmentation.

During binarization, some images were not well-binarized and the individual characters formed disconnected components. So, to make them

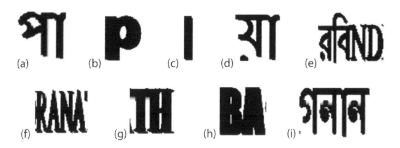

**FIGURE 3.3**
Parts (a)–(d) are well-segmented characters of a bi-scripted word, whereas (e)–(i) are under-segmented images of a bi-scripted word.

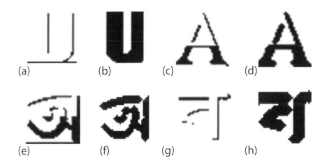

**FIGURE 3.4**
The disconnected component images are shown in (a), (c), (e), and (g); joined component images are shown in (b), (d), (f), and (h).

connected, the following component joining algorithm (algorithm 1) was used, drawing on isotropic dilation [18], a component labeling [19] and flood fill algorithm [20], and a Euclidean transform algorithm [21]. In Figure 3.4, the broken component character images and their corresponding connected component characters can be seen. Next, we thinned the images using the Zhang Suen thinning algorithm [22].

### Algorithm 1: component joining

```
Input: I, N [where I represents the binary image and
N is the number of connected components]
    1. Threshold T = 0
    2. If (N > 1) repeat steps 3 to 5 until N = 1
    3. I = EDT < = T [Where EDT is Euclidean Distance
       Transform]
    4. T = T + 0.1
    5. Update N
Output: connected component character image I.
```

## 3.4 Feature Extraction

Feature extraction is a very important phase in image recognition. In this work, we have considered two types of features, namely topology- or geometry-based and texture features.

### 3.4.1 Topology-Based Features

The topology-based [23] feature extracts the basic geometrical properties of a character image. The following features are considered in the topology of characters.

a. **Area:** Area measures the number of pixels with an intensity value I in the whole image. The area can be represented as:

$$\text{Area} = \sum_{m,n \in R} I \tag{3.1}$$

where m and n denote the x and y axes range of region R.

b. **Major Axis Length:** In an ellipse, there are two focus points (foci), a and b; the sum of the distance between these points represents the length (Figure 3.5). Major axis length can be expressed as n+p, where n and p are the distances from each focus to any random point c on the ellipse.

c. **Minor Axis Length:** Looking at Figure 3.5, the minor axis length can be expressed as $\sqrt{(n+p)^2 - m^2}$ .

d. **Eccentricity**: The meaning of eccentricity is 'off center'. That means that, as the eccentricity increases, the foci are proportionally receded from the center or off the center. It can be expressed as:

$$\text{Eccentricity} = \frac{g}{h} \tag{3.2}$$

Where g is the distance from the center c to a focus f2, and h is the distance from that focus f2 to a vertex a, which is a scalar value in the range 0 to 1. When the ratio is 0 Figure 3.6 will be a circle, and when it is 1 it will be a line.

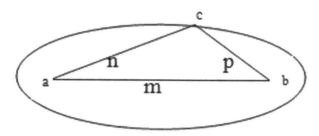

**FIGURE 3.5**
Ellipse with foci.

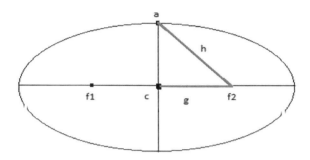

**FIGURE 3.6**
The eccentricity of an ellipse.

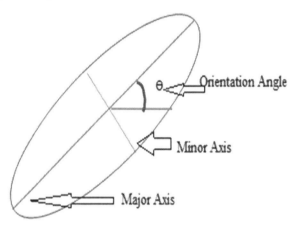

**FIGURE 3.7**
The orientation of an ellipse.

e. **Orientation:** This scalar value represents the angle between the major axis and x-axis of the ellipse that has the same second-moments as the region (Figure 3.7).

f. **Convex Area:** In the convex area no line segment between two adjacent points on the boundary ever goes outside the polygonal area and it denotes the number of pixels in the convex image.

g. **Euler Number:** The Euler number [24] is a scalar value calculated as:

$$E = C - H \tag{3.3}$$

where
    $C$   denotes the number of connected objects in the region.
    $H$   represents holes, i.e background area bounded by foreground boundary.

So, the Euler number is basically the difference between the numbers of the connected area to the number of holes.

h. **Equivalent Diameter:** The equivalent diameter can be expressed as:

$$E_d = \sqrt{\frac{(4 * \text{Area of the region})}{\pi}} \tag{3.4}$$

This scalar quantity represents the diameter of a region.

i. **Solidity:** This scalar quantity measures the ratio of the area of an object to the area of a convex hull of the object. So, the solidity of an object can be expressed as:

$$\text{Solidity} = \frac{\text{Area}}{\text{ConvexArea}} \tag{3.5}$$

j. **Extent:** Extent measures the ratio of the area of an object to the area of the bounding rectangle enclosed within or around the object, which is a scalar value:

$$\text{Extent} = \text{Area}/\text{Area of Bounding Rectangle} \tag{3.6}$$

k. **Perimeter:** Perimeter measures the distance between each neighboring pair of pixels around the border of the region; non-zero-valued pixels are part of the perimeter and are connected to at least one zero-valued pixel.

l. **Circularity:** Circularity measures the compactness of an object. It encloses the area of a given perimeter. It can be expressed as:

$$\text{Circularity} = \frac{p^2}{4 * \pi * A} \tag{3.7}$$

where P and A are the perimeter and area of the object, respectively.

## 3.4.2 Texture Feature

Texture features describe the contents of an image or a region of an image and give important low-level features and mean-oriented energy (Figure 3.8).

A discrete Gabor wavelet returns mean-oriented energy as a texture feature [25, 26]. It transforms an image g(x, y) with MXN dimensions, which can be represented as

$$W_{cd}(x,y) = \sum_a \sum_b g(x - a, y - b) \psi_{ab}{}^*(a,b) \tag{3.8}$$

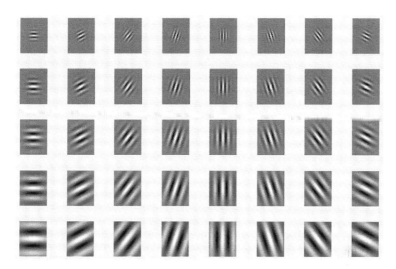

**FIGURE 3.8**
The real part of a Gabor wavelet with five scales and eight orientations.

where
    c and d   denote scale and orientations, respectively.
    a and b   are the sizes of filter mask.
    $\psi^*$   is the complex conjugate of $\psi$, where $\psi$ is a morelet and can be represented as:

$$\psi(x,y) = \left(\frac{1}{2\pi\sigma_x\sigma_y}\right)\exp\left(\frac{-1}{2}\left(\frac{x^2}{\sigma_x^2} + \frac{y^2}{\sigma_y^2}\right)\right)\exp(2\pi jwx) \qquad (3.9)$$

By using the generating function, we can have the other Gabor wavelets as follows:

$$\psi_{so}(x,y) = m^{-c}\psi(x',y') \qquad (3.10)$$

where

$$x' = r^{-c}\left(x\cos\theta + y\sin\theta\right) \text{ and } y' = \left(-x\sin\theta + y\cos\theta\right) \qquad (3.11)$$

For $m > 1$ and $\theta = \dfrac{d\pi}{t1}$

Here, t1 represents the total number of scalar of the filter.

The energy content can be considered to have scale c and direction d of image g, as:

$$E_{cd}(g) = \sum_x\sum_y \left|W_{cd}(x,y)\right| \qquad (3.12)$$

From the Gabor wavelet, 2560 features have been obtained, but a lot of them are redundant, and dealing with this large number of features degrades the performance of our work. In this regard, we have gone for dimensionality reduction by using PCA [27]. By using PCA, we considered 60 texture features and from topology-based features, 12 have been obtained. Thus, 72 features have been considered in this work.

## 3.5 Experiments

### 3.5.1 Estimation Procedure

To estimate the performance of our technique, the correct recognition rate (CRR) can be measured:

$$CRR = \frac{(\text{Number of images correctly recognised})}{(\text{Total number images})} * 100\% \qquad (3.13)$$

Along with CRR, kappa statistic and root mean square error (RMSE) metrics also play a vital role in recognition, defined as follows:

$$Kappa = \frac{(\text{observed accuracy} - \text{expected accuracy})}{(1 - \text{expected accuracy})} \qquad (3.14)$$

$$RMSE = \sqrt{\frac{\sum_{m=1}^{S}\sum_{n=1}^{T}\left[f(m,n) - g(m,n)\right]^2}{ST}} \qquad (3.15)$$

Where f(x,y) and g(x,y) represent reference and target images, respectively with sizes S, X, T.

Here, we have used five-fold cross-validation to estimate the performance of our technique. In the five-fold process, the whole dataset has been bifurcated into five equally sized parts, one of which has been considered a testing part, and the remainder as a training set; the average of all the results produces the final result.

### 3.5.2 Results and Analysis

We tested our method using different classifiers like SVM [28], RBF [29], random forest [30], and MLP [31], and in Table 3.1 the different parameter values of their corresponding classifiers are depicted.

**TABLE 3.1**

Classifiers with Different Parameters

| Classifier | Root Mean Square Error | Kappa Statistic | CRR(%) |
|---|---|---|---|
| SVM | 0.2210 | 0.8419 | 71.29 |
| RBF | 0.2155 | 0.8685 | 91.90 |
| Random Forest | 0.2790 | 0.8771 | 93.90 |
| MLP | 0.2894 | 0.8852 | 94.95 |

**TABLE 3.2**

Confusion Matrix for the Highest Accuracy Rate of Bi-Script Identification Using Random Forest

| | Bangla | Devanagari | Roman |
|---|---|---|---|
| Bangla | 1,554 | 45 | 21 |
| Devanagari | 14 | 123 | 7 |
| Roman | 34 | 12 | 818 |

It can be observed from Table 3.1 that MLP gives the highest accuracy (94.95%) after random forest (93.90%), RBF (91.90%), and SVM (71.29%). The Kappa Statistic and RMSE of MLP are 0.8852 and 0.2894, respectively. In MLP, the parameter learning rate has been taken as 0.3 and momentum as 0.2, and it has been found nine numbers of hidden layer gives the best result.

In Table 3.2, we tabulate the confusion matrix of the highest performing classifier on our dataset.

In Table 3.2, it is observed that some characters are not correctly recognized as the corresponding class: 1,554 characters are correctly classified as Bangla, but 45 and 21 characters are classified as Devanagari and Roman, respectively. This misclassification occurs due to similarity in character geometry and binarization issues, i.e. some character shapes vanish after binarization.

## 3.6 Conclusion

Artistic multi-character script recognition is a very new problem of script identification and document processing [32–35] work. Aspects like size, color, texture, complex background, and skewness make this problem challenging to researchers. From our technique, 94.95% accuracy has been observed, but this rate can be increased by using the preprocessing phase more accurately and with other feature extraction algorithms along with the above-described methods. The database creation is another issue and collecting more natural images is our next plan.

## References

1. Ghosh, D., Dube, T., & Shivaprasad, A. Script recognition—A review. *IEEE Transactions on Pattern Analysis and Machine Intelligence*, 32(12), 2142–2161 (2010).

2. Obaidullah, S. M., Bose, A., Mukherjee, H., Santosh, K. C., Das, N., & Roy, K. Extreme learning machine for handwritten Indic script identi_cation in multi-script documents. *Journal of Electronic Imaging*, 27(5), 051214 (2018).

3. Obaidullah, S. M., Santosh, K. C., Halder, C., Das, N., & Roy, K. Automatic Indic script identification from handwritten documents: Page, block, line, and word-level approach. *International Journal of Machine Learning and Cybernetics*, 10(1), 1–20 (2019).

4. Obaidullah, S. M., Halder, C., Santosh, K. C., Das, N., & Roy, K. PHDIndic_11: Page-level handwritten document image dataset of 11 official Indic scripts for script identification. *Multimedia Tools and Applications*, 77(2), 1643–1678 (2018).

5. Obaidullah, S. M., Goswami, C., Santosh, K. C., Das, N., Halder, C., & Roy, K. Separating Indic scripts with matra for effective handwritten script identification in multi-script documents. *International Journal of Pattern Recognition and Artificial Intelligence*, 31(05), 1753003 (2017).

6. Obaidullah, S. K., Santosh, K. C., Halder, C., Das, N., & Roy, K. Word-level multi-script Indic document image dataset and baseline results on script identification. *International Journal of Computer Vision and Image Processing (IJCVIP)*, 7(2), 81–94 (2017).

7. Obaidullah, S. M., Santosh, K. C., Das, N., Halder, C., & Roy, K. Handwritten Indic script identification in multi-script document images: A survey. *International Journal of Pattern Recognition and Artificial Intelligence*, 32(10), 1856012 (2018).

8. Hangarge, M., Santosh, K. C., & Pardeshi, R. Directional discrete cosine transform for handwritten script identification. In: *2013 12th International Conference on Document Analysis and Recognition* (pp. 344–348). IEEE (2013).

9. Mori, S., Suen, C. Y., & Yamamoto, K. Historical review of OCR research and development. *Proceedings of the IEEE*, 80(7), 1029–1058 (1992).

10. Tan, T. N. Rotation invariant texture features and their use in automatic script identification. *IEEE Transactions on Pattern Analysis and Machine Intelligence*, 20(7), 751–756 (1998).

11. Chanda, S., Pal, S., Franke, K., & Pal, U. Two-stage approach for word-wise script identification. In: *10th International Conference on Document Analysis and Recognition, 2009. ICDAR'09* (pp. 926–930). IEEE (2009).

12. Padma, M. C., & Vijaya, P. A. Identification of Telugu, devanagari and English scripts using discriminating. *Journal of Computer Science*, 1, 64–78 (2009).

13. Manjula, S., & Hegadi, R. S. Identification and classification of multilingual document using maximized mutual information. In: *2017 International Conference on Energy, Communication, Data Analytics and Soft Computing (ICECDS)* (pp. 1679–1682). IEEE (2017).

14. Busch, A., Boles, W. W., & Sridharan, S. Texture for script identification. *IEEE Transactions on Pattern Analysis and Machine Intelligence*, 27(11), 1720–1732 (2005).

15. Pal, U., & Chaudhuri, B. B. Automatic identification of English, Chinese, Arabic. In: *Proceedings of Sixth International Conference on Devnagari and Bangla Script Line. In Document Analysis and Recognition, 2001* (pp. 790–794). IEEE (2001)

16. Otsu, N. A threshold selection method from gray-level histograms. *IEEE Transactions on Systems, Man, and Cybernetics*, 9(1), 62–66 (1979).

17. Sridevi, N. *Optimized Complex Extreme Learning Machine for Classification of 11th Century Handwritten Tamil Scripts* (2013).

18. Vincent, L. Morphological algorithms. In: E. Dougherty, ed., *Mathematical Morphology in Image Processing*, Marcel Dekker, New York, 255–258 (1992).

19. Samet, H., & Tamminen, M. Efficient component labeling of images of arbitrary dimension represented by linear bintrees. *IEEE Transactions on Pattern Analysis and Machine Intelligence*. IEEE. 10(4), 579 (1988).

20. Silvela, J., & Portillo, J. Breadth-first search and its application to image processingproblems. *IEEE Transactions on Image Processing*, 10(8), 1194–1199 (2001).

21. Breu, H., Gil, J., Kirkpatrick, D., & Werman, M. Linear time Euclidean distance transform algorithms. *IEEE Transactions on Pattern Analysis and Machine Intelligence*, 17(5), 529–533.

22. Zhang, T. Y., & Suen, C. Y. A fast parallel algorithm for thinning digital patterns. *Communications of the ACM*, 27(3), 236–239 (1984).

23. Yang, M., Kpalma, K., & Ronsin, J. *A Survey of Shape Feature Extraction Techniques, Pattern Recognition*, Peng-Yeng Yin (Ed.) (pp. 43–90) (2008).

24. Lin, X., Sha, Y., Ji, J., & Wang, Y. A proof of image Euler number formula. *Science in China, Series F*, 49(3), 364–371 (2006).

25. Ma, W. Y., & Manjunath, B. S. Texture features and learning similarity. In: *Proceedings CVPR'96 IEEE Computer Society Conference on Computer Vision and Pattern Recognition* (pp. 425–430). IEEE (1996).

26. Zhang, D., Wong, A., Indrawan, M., & Lu, G. Content-based image retrieval using Gabor texture features. *IEEE Transactions PAMI*, 13–15 (2000).

27. Jolliffie, I. T. *Principal Component Analysis, Series*. Springer Series in Statistics, 2nd ed., Springer, New York, (2002), XXIX, 487 p. 28 illus.

28. Cusano, C., Ciocca, G., & Schettini, R. Image annotation using SVM. In: *Internet Imaging V* (Vol. 5304, pp. 330–339). International Society for Optics and Photonics (2003).

29. Er, M. J., Wu, S., Lu, J., & Toh, H. L. Face recognition with radial basis function (RBF) neural networks. *IEEE Transactions on Neural Networks*, 13(3), 697–710 (2002).

30. Rodriguez-Galiano, V. F., Ghimire, B., Rogan, J., Chica-Olmo, M., & Rigol Sanchez, J. P. An assessment of the effectiveness of a random forest classifier for land-cover classification. *ISPRS Journal of Photogrammetry and Remote Sensing*, 67, 93–104 (2012).

31. Alkan, A., Koklukaya, E., & Subasi, A. Automatic seizure detection in EEG using logistic regression and artificial neural network. *Journal of Neuroscience Methods*, 148(2), 167–176 (2005).

32. Santosh K. C. g-DICE: Graph mining-based document information content exploitation. *International Journal on Document Analysis and Recognition (IJDAR)*, 18(4), 337–355 (2015).

33. Santosh, K. C., & Belaïd, A. Document information extraction and its evaluation based on client's relevance. In: *2013 12th International Conference on Document Analysis and Recognition* (pp. 35–39). IEEE (2013).

34. Bouguelia, M. R., Nowaczyk, S., Santosh, K. C., & Verikas, A. Agreeing to disagree: Active learning with noisy labels without crowdsourcing. *International Journal of Machine Learning and Cybernetics*, 9(8), 1307–1319 (2018).
35. Vajda, S., & Santosh, K. C. A fast k-nearest neighbor classifier using unsupervised clustering. In: *International Conference on Recent Trends in Image Processing and Pattern Recognition* (pp. 185–193). Springer, Singapore (2016).

# 4

# A Study on the Extreme Learning Machine and Its Applications

Himadri Mukherjee, Sahana Das, Subhashmita Ghosh,
Sk Md Obaidullah, KC Santosh, Nibaran Das and Kaushik Roy

## CONTENTS

## 4.1 Introduction

The primary issue for the establishment of extreme learning machines (ELMs) is the slow learning rate that occurs in neural networks. The ELM algorithm is near the single hidden layer feed forward neural network (SHLFFNN) for regression as well as classification with three prime differences. Those are as follows:

- The number of neural network hidden neurons which use back propagation algorithm for training is much less than the number of ELM hidden neurons.

- By utilizing the values from sustained uniform distribution, the weights are randomly created from input to hidden layer, called 'random projection'.
- To solve the output weight, the least square error regression is used, i.e. the Moore–Penrose generalized algorithm.

ELMs are the contemporary flavor of the random projection, accompanied by some hypothetical properties which can potentially fit any structure along with a least square procedure. The concept of this learning machine lies on the factual theory of risk minimization and only one iteration is needed to finish the learning process.

Initially, Huang et al. [1] proposed ELMs to address the complication caused by gradient descent-based algorithms. But in the 1940s, McCulloch and Pitts [2] gave the logical view that unlocked the path of research of artificial neural networks (ANNs).

The ELM was used to explain a linear equation when the weights of hidden to output layers are projected. Moore–Penrose's generalized inverse [3] is applied to solve the general system which led to solving the linear equation. The ELM has been used to solve classification and regression problems [4, 5] and has produced promising results in several instances.

In this chapter we talk about the preliminaries, extensions, applications and issues of ELMs.

---

## 4.2 Preliminaries

Assuming that a SHLFFNN with L number of hidden neurons, where $G\ (w_i, b_i, x)$ is the resultant output of the hidden layer node, and with $b_i$ bias corresponding to x input vector with the weight $w_i\ (w_1, w_2, ..., w_i)$ connected to hidden to output layer, then $\beta_i\ (\beta_1, \beta_2, \beta_i)$ is the output function f ( ) given by:

$$f(x) = \sum_{i=1}^{L} \beta_i G(w_i, b_i, x_i) \tag{4.1}$$

G( ) will compute for additive hidden neurons as:

$$G(w_i, b_i, x_i) = g(w_i.x + b_i) \tag{4.2}$$

Here, the activation function is represented by $g: R \rightarrow R$.

Equations 4.1 and 4.2 aid in the generation of the following:

$$H\beta = Y \tag{4.3}$$

$$\beta = H^{-1}Y \tag{4.4}$$

where the hidden layer output matrix is $H$.

$$H = \begin{bmatrix} G(w_1, b_1, x_1) & \cdots\cdots & G(w_L, b_L, x_1) \\ G(w_1, b_1, x_M) & \cdots\cdots & G(w_L, b_L, x_M) \end{bmatrix}$$

$$B = \begin{bmatrix} \beta_1 \\ \vdots \\ \beta_L \end{bmatrix}$$

$$Y = \begin{bmatrix} Y_1 \\ \vdots \\ Y_M \end{bmatrix}$$

where

$\{(x_i, y_i)\}_{i=1,2,3,\cdots M}$ is the training set.

$x_i = (x_1, x_2, \ldots, x_M)$ is supposed to be the input vector.

$y_i = (y_1, y_2, \ldots, y_k)$ is the output vector.

For a training set $(x_i, y_i)$ with the count of hidden neurons $L$ and the activation or transfer function $g(\ )$, the algorithm of ELM is represented in three steps.

1. Randomly assign the input weight $w_i$ and bias $b_i$ for the hidden layer.
2. Determine the hidden layer to output layer $G(w_i, b_i, x)$.
3. Lastly, calculate $\beta$ according to the previously mentioned Equations 4.3 and 4.4.

---

## 4.3 Activation Functions of ELM

The activation function is one of the most formidable features of an ANN, a single layer feedforward neural network for learning and justifying really complex material. The main purpose of the activation function is to transfer the input neuron to the output signal. Now, the question is, what if we don't use this function? Without it, the output function is just a simply linear function, and the ANN becomes just a linear regression model that is not able to learn any complicated data set.

Mostly, ELM deals with the 'sigmoid', 'sine', 'hardlimit', 'tribas' and 'radbas' activations. The following is a brief discussion of these.

### 4.3.1 Sigmoid Function

A sigmoid activation function produces an 'S' formation curve, which ranges between 0 to 1. This function is widely used and easy to understand, but there are some major aspects that have made the sigmoid fall out of popularity. The issues result in this function being 0-centered. It saturates and kills the gradient and also has a slow convergence. The equation is as follows:

$$g(w,b,x) = \frac{1}{1+\exp(-w.x+b)} \tag{4.5}$$

### 4.3.2 Hardlimit Function ('Hardlim')

The activation function hardlimit, written as 'hardlim', insists on a node to output 1 if it reaches the threshold value, or else it is set to 0. This kind of activation function is mainly used with perceptron learning. The output of this transfer function is strictly either 0 or 1. The equation is as follows:

$$g(w,b,x) = 1 \quad \text{If } (w.x-b) \geq 0$$
$$g(w,b,x) = 0 \quad \text{If } (w.x-b) \leq 0 \tag{4.6}$$

### 4.3.3 Radial Basis Function ('Radbas')

The radbas activation function in an ELM is a real-valued activation function. It is defined only on Euclidean space. The aggregation of a radial basis function is mainly applied to the function approximation problem. ELM-radbas functions use a Gaussian kernel function. For this function, the output can be ranged between –1 to +1.

The equation is as follows:

$$g(w,b,x) = 1 \exp\left(-b\|x-w\|^2\right) \tag{4.7}$$

### 4.3.4 Sine Function

The sine activation, known as the 'sine C activation', is a very popular function in ELMs for illustrating classifications and regressions. For this saturating activation function, the output for negative and large positive inputs converges to 0. This function is defined as follows:

$$y = \sin(x)/x \tag{4.8}$$

As we know, dividing a value to 0 makes an undefined equation. So, an exception to the above function, where x is equal to 0, SinC (0), is noted. The value of 1 is an exception.

$$y = \frac{\sin(x)}{x} = 0 \quad \text{for } x \neq 0$$

$$y = \frac{\sin(x)}{x} = 1 \quad \text{for } x = 0$$

(4.9)

### 4.3.5 Triangular Basis Function ('Tribas')

As the graphical shape of the triangular basis function is like a triangle, it is also known as 'hat', 'triangle' and 'tent function'. To derive realistic signals for this function, mainly a kernel integral transform function is used. The following equation is derived from this transfer function.

$$g(w, b, x) = 1 \quad \text{if } -1 \leq x \leq 1$$
$$g(w, b, x) = 0 \quad \text{otherwise}$$

(4.10)

## 4.4 Metamorphosis of an ELM

Due to some boundaries and pitfalls of ELMs, the idea of metamorphosis of an ELM is established. Some of the metamorphosis is briefly implemented.

- An incremental ELM or I-ELM [6] is proposed for the handling of the complication occurring during the learning procedure. Here, the hidden nodes are randomly allocated to the hidden layer one by one, and the output weights are also coagulated. As the transfer function of ELMs is continuous in nature, the I-ELM is one of the most muscular extensions for fixing any expected function in the compound domain.

- An evolutionary ELM, also known as an E-ELM [7], is enhanced to determine the weight without tuning to the least square error minimization. Within the models of a conventional ELM, there exist a vast number of hidden nodes, and thus the input weight determination and network generalization performance are not satisfactory. For this reason, the concept of E-ELM has been established. The resultant output of the E-ELM is better than the conventional algorithm.

- To solve the benchmark problem of classification, a pruning ELM or P-ELM [8] is proposed. It is a very popular method for resolving regression and classification problems. A P-ELM can deal with more than just the over-fitting problem. First it gathers the nodes which are not suitable for the network, then diminishes the remaining ones from the network.

- A voting-based ELM, known as V-ELM [9], reduces misclassification problems. For the Classical ELM, the data close to the classification boundary may be misclassified due to the fluent abstraction of bias and input weights. By applying a major voting mechanism after multiple training, the resulting V-ELM deals with such misclassification problems.

- An online sequential ELM or OS-ELM [10] may or may not set the data size for learning. In an OS-ELM, the weights of input towards the hidden layer are rapidly assembled, but the output weights are regulated in a refined manner. It is clearly discerned that the OS-ELM is swifter and promotes better generalization enactment.

- To resolve tuning parameters as well as local minimum problems, Li et al. [11] proposed a fully complexed ELM or C-ELM. It can move comfortably in any complex domain, and its learning phase is significantly high-speed and acquires symbol error rate (SER).

- To suppress computational complexity, an error minimized ELM or EM-ELM [12] is proposed. Beyond this metamorphosis, other extended versions of ELMs are proposed these days.

Not only are these extensions of an ELM implemented, but also other metamorphoses like the O-ELM, Fuzzy ELM, Weighted ELM, Regularized ELM, Robust ELM, Parallel ELM, TLS ELM, S-ELM and others. These are extended to handle several complications that occur in ELMs for the classification, regression, close boundary, etc., problems.

## 4.5 Applications of ELMs

ELMs as well as their extensions may contribute to many fields. ELMs may serve different aspects of classification as well as regression problems. Some domains where the applications of ELMs are highly popular are discussed below.

### 4.5.1 ELMs in Document Analysis

Obaidullah et al. [13] have experimented with different activation functions in the task of script identification from images of handwritten documents. Extreme learning machines were vividly used to recognize patterns. Chacko et al. [14] have used ELMs for handwritten character identification along with wavelet energies to recognize 3D objects. Mahmoud et al. [15] have used ELMs combined with SVMs for recognizing offline handwritten Arabic numerals.

### 4.5.2 ELMs in Medicine

Machine learning has had remarkable contributions in the fields of medical [16–21] and health care. ELMs have been used to recognize thyroid stimulating hormones by Li et al. [22]. Daliri et al. [23] showed that ELMs can also detect lung cancer. Detecting heart problems using an ELM has also been shown, by Karpagachelvi et al. [24].

### 4.5.3 ELM in Audio Signal Processing

ELMs have proved to be very beneficial in diverse audio-based applications. Mukherjee et al. [25] have demonstrated the use of ELMs in musical instruments as well as instrument family identification from audio signals in the presence of different types of noises. In another instance, they used an ELM for voice activity detection [26]. Voice activity detection is a very important step for processing speech signals, detecting segments with vocal data.

### 4.5.4 ELM in Other Pattern Recognition Problems

Mohammed et al. [27] defined how an ELM can recognize a human face for 2D bidirectional components [28]. ELMs may also be able to recognize multi-label faces, as implemented by Zong et al. [29]. ELMs have also been used for regression analysis. As per He et al. [30], a Parallel ELM can interpret regression on vast scale data. Li et al. [31] proposed a ridge regression to conquer the consequence of unsettled data. Balasundaram [32] proposed an E-incentive regression model for neuro computing and its applications.

So, ELMs are used in broad aspects and different areas. Their applications are not limited to image processing but are also useful in medicine, retail and other domains.

## 4.6 Challenges of ELM

Though extreme learning machines are very popular in many areas, some issues clearly remain. One of the main problems is that we can't get any optimal solutions from ELMs. ELMs cannot play any role in parallel computing or distributed computing. Compared to learning algorithms, the performance is quite unstable. To resolve the classification problem and other boundaries, ELMs need to be tested more.

## 4.7 Conclusion

In this chapter we discussed the architecture of ELMs followed by their algorithms, types of activation function, extensions and applications. We talked about the popularity of ELMs and also the pitfalls which are present in different variants of the same. To overcome the pitfalls of ELMs, some extensions were introduced. ELMs are a very specified approach to pattern reorganization, technologies and other areas, but the open problems present further tasks for researchers.

## References

1. Huang, G. B., Zhu, Q. Y., & Siew, C. K. (2004, July). Extreme learning machine: A new learning scheme of feed forward neural networks. In: *Proceedings of the 2004 IEEE International Joint Conference on Neural Networks, 2004* (Vol. 2, pp. 985–990). IEEE.
2. McCulloch, W. S., & Pitts, W. (1943). A logical calculus of the ideas immanent in nervous activity. *The Bulletin of Mathematical Biophysics*, 5(4), pp. 115–133.
3. Rao, C. R. (1971). *Generalized Inverse of Matrices and Its Applications*. John Wiley & Sons, New York.
4. Rong, H. J., Ong, Y. S., Tan, A. H., & Zhu, Z. (2008). A fast pruned-extreme learning machine for classification problem. *Neurocomputing*, 72(1–3), 359–366.
5. Huang, G. B., Zhou, H., Ding, X., & Zhang, R. (2012). Extreme learning machine for regression and multiclass classification. *IEEE Transactions on Systems, Man, and Cybernetics, Part B (Cybernetics)*, 42(2), 513–529.
6. Huang, G. B., Chen, L., & Siew, C. K. (2006). Universal approximation using incremental constructive feedforward networks with random hidden nodes. *IEEE Transactions on Neural Networks*, 17(4), 879–892.
7. Zhu, Q. Y., Qin, A. K., Suganthan, P. N., & Huang, G. B. (2005). Evolutionary extreme learning machine. *Pattern Recognition*, 38(10), 1759–1763.
8. Rong, H. J., Ong, Y. S., Tan, A. H., & Zhu, Z. (2008). A fast pruned-extreme learning machine for classification problem. *Neurocomputing*, 72(1–3), 359–366.
9. Cao, J., Lin, Z., Huang, G. B., & Liu, N. (2012). Voting based extreme learning machine. *Information Sciences*, 185(1), 66–77.
10. Liang, N. Y., Huang, G. B., Saratchandran, P., & Sundararajan, N. (2006). A fast and accurate online sequential learning algorithm for feedforward networks. *IEEE Transactions on Neural Networks*, 17(6), 1411–1423.
11. Li, M. B., Huang, G. B., Saratchandran, P., & Sundararajan, N. (2005). Fully complex extreme learning machine. *Neurocomputing*, 68, 306–314.
12. Feng, G., Huang, G. B., Lin, Q., & Gay, R. K. L. (2009). Error minimized extreme learning machine with growth of hidden nodes and incremental learning. *IEEE Transactions on Neural Networks*, 20(8), 1352–1357.
13. Obaidullah, S. M., Bose, A., Mukherjee, H., Santosh, K. C., Das, N., & Roy, K. (2018). Extreme learning machine for handwritten Indic script identification in multiscript documents. *Journal of Electronic Imaging*, 27(5), 051214.

14. Chacko, B. P., Krishnan, V. V., Raju, G., & Anto, P. B. (2012). Handwritten character recognition using wavelet energy and extreme learning machine. *International Journal of Machine Learning and Cybernetics*, 3(2), 149–161.

15. Mahmoud, S. A., & Olatunji, S. O. (2009). Automatic recognition of off-line handwritten Arabic (Indian) numerals using support vector and extreme learning machines. *International Journal of Imaging*, 2(A09), 34–53.

16. Ruikar, D. D., Santosh, K. C., & Hegadi, R. S. (2019). Automated fractured bone segmentation and labeling from CT images. *Journal of Medical Systems*, 43(3), 60.

17. Vajda, S., Karargyris, A., Jaeger, S., Santosh, K. C., Candemir, S., Xue, Z., Antani, S., & Thoma, G. (2018). Feature selection for automatic tuberculosis screening in frontal chest radiographs. *Journal of Medical Systems*, 42(8), 146.

18. Ruikar, D. D., Hegadi, R. S., & Santosh, K. C. (2018). A systematic review on orthopedic simulators for psycho-motor skill and surgical procedure training. *Journal of Medical Systems*, 42(9), 168.

19. Santosh, K. C., & Wendling, L. (2018). Angular relational signature-based chest radiograph image view classification. *Medical and Biological Engineering and Computing*, 1–12.

20. Santosh, K. C., & Antani, S. (2018). Automated chest x-ray screening: Can lung region symmetry help detect pulmonary abnormalities? *IEEE Transactions on Medical Imaging*, 37(5), 1168–1177.

21. Zohora, F. T., Antani, S., & Santosh, K. C. (2018, March). Circle-like foreign element detection in chest x-rays using normalized cross-correlation and unsupervised clustering. In: *Medical Imaging 2018: Image Processing* (Vol. 10574, p. 105741V). International Society for Optics and Photonics.

22. Li, L. N., Ouyang, J. H., Chen, H. L., & Liu, D. Y. (2012). A computer aided diagnosis system for thyroid disease using extreme learning machine. *Journal of Medical Systems*, 36(5), 3327–3337.

23. Daliri, M. R. (2012). A hybrid automatic system for the diagnosis of lung cancer based on genetic algorithm and fuzzy extreme learning machines. *Journal of Medical Systems*, 36(2), 1001–1005.

24. Karpagachelvi, S., Arthanari, M., & Sivakumar, M. (2012). Classification of electrocardiogram signals with support vector machines and extreme learning machine. *Neural Computing and Applications*, 21(6), 1331–1339.

25. Mukherjee, H., Obaidullah, S. M., Phadikar, S., & Roy, K. (2018). MISNA-A musical instrument segregation system from noisy audio with LPCC-S features and extreme learning. *Multimedia Tools and Applications*, 1–26.

26. Mukherjee, H., Obaidullah, S. M., Santosh, K. C., Phadikar, S., & Roy, K. (2018). Line spectral frequency-based features and extreme learning machine for voice activity detection from audio signal. *International Journal of Speech Technology*, 1–8.

27. Mohammed, A. A., Minhas, R., Wu, Q. J., & Sid-Ahmed, M. A. (2011). Human face recognition based on multidimensional PCA and extreme learning machine. *Pattern Recognition*, 44(10–11), 2588–2597.

28. Wang, H., Hussain, M. F., Mukherjee, H., Obaidullah, S. M., Hegadi, R. S., Roy, K., & Santosh, K. C. (2018). An empirical study: ELM in face matching. In: *International Conference on Recent Trends in Image Processing & Pattern Recognition* (pp. 277–287). Springer, The Netherlands.

29. Zong, W., & Huang, G. B. (2011). Face recognition based on extreme learning machine. *Neurocomputing*, 74(16), 2541–2551.

30. He, Q., Shang, T., Zhuang, F., & Shi, Z. (2013). Parallel extreme learning machine for regression based on MapReduce. *Neurocomputing*, 102, 52–58.
31. Li, G., & Niu, P. (2013). An enhanced extreme learning machine based on ridge regression for regression. *Neural Computing and Applications*, 22(3–4), 803–810.
32. Balasundaram, S. (2013). On extreme learning machine for ε-insensitive regression in the primal by Newton method. *Neural Computing and Applications*, 22(3–4), 559–567.

# 5

# A Graph-Based Text Classification Model for Web Text Documents

Ankita Dhar, Niladri Sekhar Dash and Kaushik Roy

## CONTENTS

## 5.1 Introduction

A significant mass of accessible digital data is found to be in unstructured or semi-structured form. Thus, for a user to understand the importance and applicability of the information in this form according to their needs, and to access that information efficiently, there is a pressing need for automated information extraction and text document classification/categorization systems. However, the task is not very simple, as text data is formed in natural language. Communication in the form of writing and speaking in a language desires an understanding of syntactic and semantic relationships along with the enriching perceptions within that language. In spite of these complexities, most studies in this domain of text mining cast light on one particular facet of a language, that is, the frequency of a term. Representation of text documents using the bag-of-words approach is believed to be self-sufficient without expressing the syntactic or semantic relationships between terms or series of terms. But these become important while categorizing short text documents where capturing the semantic relationships among terms is necessary for identification, because of the usage of synonyms or interconnected terms in the texts despite similar casing categories.

In the present experiment, a graph-based text classification approach has been designed, involving syntactic and semantic relationships along with term frequencies, to enhance the performance of the text categorization system. The experiment was carried out on a dataset of 9000 Bangla text documents covering eight different text categories acquired from online news corpora, various magazines and webpages. Sets of text documents are treated as graph sets to which an algorithm of a weighted graph is executed to extract subgraphs which can be further used to generate the feature vectors from the considered dataset for classification. A weighted graph algorithm is taken into account for extracting the most applicable subgraphs, which successively increases the rate of classification and computational efficiency. Contrary to the traditional vector space model and bag-of-words approaches, graph-based models perform more efficiently by considering the structural information of the text documents. Graph-based models are becoming an alternative way to represent text because of the ability to encapsulate important facts regarding texts, such as ordering a term, co-occurrence of terms and relationships among different terms. Furthermore, compared to the existing methods applied to Bangla text categorization by various researchers, the graph-based learning method shows the advantage of involving related knowledge.

The paper is organized as follows. In Section 5.2, the basis of text categorization and various machine learning algorithms is provided by surveying literature in this domain. Section 5.3 provides detailed insights about the data collection, approaches used for pre-processing, graph generations for the extraction of features and the classification algorithms. Section 5.4 explores the results and other perceptions obtained by the graph-based methods. Besides analyzing the outcomes and potential developments, Section 5.4 also provides a comparison among several commonly used classification algorithms as well as a comparison with existing methods. In Section 5.5, the paper is concluded showing some future directions.

## 5.2 Related Works

### 5.2.1 English

From the literature survey it can be observed that text categorization in English has gained great attention by researchers. A few works are briefly discussed here. For instance, Dasondi et al. [1] proposed a new text categorization technique for classifying text documents from social media. The proposed approach analyzes the text patterns automatically and extracts their positive and negative directions based on the analyzed patterns. The approach employs the possibility of term occurrence, generation of a sentence and the assigning of weights to the terms for observing the location of the text data. They formulated their approach in such a way that various combinations of terms can be observed using a graphical model. The estimated possibilities and the assigned weights are used for generating the directed weighted graphs, and based on this graph the text documents were classified. Guru et al. [2] introduced an alternative scheme for choosing the most appropriate subset of the primary feature set for text classification. Based on a feature set and standard feature selection methods, the proposed methodology assigns rank to the features in a cluster rather assigning the ranks individually. According to their study, a cluster of features with $n^{th}$ rank is much more dominant than the cluster of features with $(n + 1)^{th}$ rank. Thus, various clusters of features were developed that have the ability to distinguish one category from another. The advantageous factor of their technique is its capability of automatically discarding the relevant features while ranking the clusters of features. Also, the approach efficiently handles the overlying categories based on the choice of low-ranked yet dominant features. The experiment was also expanded on three standard datasets using four commonly used classification algorithms such as the support vector machine (SVM) and the naive Bayes (NB) to observe the performance of the system over the existing methods. Asim et al. [3] provide a comparative study of the nine widely used feature selection approaches on six standard datasets, namely K1a, RE1, WAP, Enron, RE0 and Spam, among which K1a and RE0 are highly skewed datasets using NB and SVM algorithms. Koswari et al. [4] proposed a technique called Random Multimodel Deep Learning (RMDL) that solves the issue of perceiving the ideal deep learning network by enhancing the accuracy through ensembles of deep learning networks. They experimented in the IMDB dataset and obtained a maximum accuracy of 90.79% using a 15 RMDL model, and also observed that RMDL performs way better in comparison to the traditional approaches. Feng et al. [5] presented a new scheme for weighting the terms based on a probabilistic model using a matching score function and observed the effectiveness of the methods using KNN and SVM on small and large texts respectively.

### 5.2.2 Chinese, Japanese and Persian

Besides English, various works that have been carried out for Chinese, Japanese and Persian can be referred to here. To name a few, Wang and Liu [6] introduced a novel technique for the categorization of Chinese text documents using graph-based KNN. They reduced the feature dimensions using a feature selection method and presented an improved graph-based text classification method using a KNN algorithm to assign the particular domain to the text documents in the training set. From their results, it can be observed that the graph-based model performs better compared to the traditional VSM-based KNN approach with respect to accuracy and execution time. Wei et al. [7] conducted the experiments over Chinese corpora consisting of 14,150 Chinese text documents from 12 classes based on three parameters: (i) comparison of the performance among absolute and relative text frequency and N-gram frequency; (ii) comparison between sparseness and feature correlation; and (iii) performance comparison among 0/1 logical value, N-gram frequency numeric value and TF-IDF value. Wenliang et al. [8] used the NEU TC dataset containing 14,459 texts from 37 domains and applied document frequency and chi square statistics for developing the class distribution of tokens. They have also used a Global Clustering Model, referred to as a global CM, for Chinese text categorization. Mikawa et al. [9] proposed an efficient approach using extended cosine measure for classification of Mainichi newspapers and customer reviews in Japanese. The approach measures the distance between two documents based on vector space model. In the case of the Persian language, Parvin et al. [10] classified Persian text documents from five text categories by using normalized mutual information along with a thesaurus and obtained a maximum accuracy of 81.45% for the value of K = 1.

### 5.2.3 Arabic and Urdu

Besides the cases of English, Chinese, Japanese and Persian, several techniques of text categorization have also been explored for the Arabic language. To mention a few, Al-Tahrawi [11] classified 1500 news articles from the Al Jazeera Arabic News corpus that were uniformly scattered into five domains using the chi square method. Al-Radaideh and Al-Khateeb [12] proposed a classification algorithm based on associative rules for classifying Arabic medical text documents; Haralambous et al. [13] implemented light stemming and rootification procedures along with the TF-IDF and dependency grammar features on the obtained dataset from the Kalimat corpus; Ali and Ijaz [14] provided a comparative study of the statistical methods for classifying Urdu text documents using NB and SVM. Mainly they pre-processed the documents using language-specific techniques for developing a standard and less dimensional feature vector. From the experiments, it can be seen that the performance increases when the value of n increases up to 3, but decreases when the values reaches 4.

### 5.2.4 Indian Languages except Bangla

Recent research carried out on text classification in the Indian languages are briefly discussed here. For instance, Gupta and Gupta [15] combined NB- and ontology-based techniques for classifying 184 Punjabi text documents extracted from various news corpora that covered seven sub-domains of sports. Their experiment has shown that the combined approach performs better compared to other approaches. For classification of Tamil text documents, ArunaDevi and Saveeth [16] used the C-feature extraction technique, which takes pairs of terms into consideration. For Marathi texts, Patil and Bogiri [17] performed automatic Marathi text document categorization on 200 documents of 20 categories based on user profiles using the label induction grouping (LINGO) algorithm. In another experiment, Patil and Game [18] used four algorithms among which NB proved to be more effective in accuracy and time for classifying Marathi text documents. Classification of texts using a language-specific WordNet for Assamese and Sanskrit were performed by Sarmah et al. [19] and Mohanty et al. [20]. In the case of Assamese [19], four classes were considered and an encouraging result of 90.27% accuracy has been obtained for the same.

### 5.2.5 Bangla

The state of evolution of text classification approaches is in a sorry state for Bangla. The lack of information for this field shows that an extended journey is waiting for us before we notice the urgent need for more research in this language. Some of the works performed in this language are reported here. Mansur et al. [21] applied the N-gram method for text document classification including only one year of data from the Pratham Alo of Bangla newspaper text corpus. In another experiment, Mandal and Sen [22] experimented with four supervised learning methods for classifying 1000 labeled documents comprised of 22,218 tokens into five categories. They achieved a maximum accuracy of 89.14% with the SVM method. On the other hand, a stochastic gradient descent (SGD) classifier was used by Kabir et al. [23] to classify text documents of nine text categories and achieved an accuracy of 93.85%, which is better compared to the previous one. In another work [24], an accuracy of 92.57% was observed based on TF-IDF and SVM algorithms for classifying Bangla text documents of 12 text domains. Cosine similarity and Euclidean distance algorithms were used by Dhar et al. [25] to categorize 1000 web text documents from five domains: Business, State, Medical, Sports and Science based on a TF-IDF weighting scheme; they attained a maximum accuracy of 95.80% for the cosine similarity measure. In another study, a new scheme of reduction technique was proposed by Dhar et al. [26] in which the top 40% of terms were taken into consideration from the total set of tokens extracted from the training documents using TF-IDF approach. The experiment was conducted on 1960 text documents from five domains and attained encouraging outcomes using a LIBLINEAR classifier. The brief survey discussed here surely shows that there is a pressing need to develop automatic text categorization systems at least for Bangla language users to serve their purposes in a more refined and efficient manner.

## 5.3 Proposed Methodology

An overview of the proposed methodology is illustrated in Figure 5.1. On a feature set acquired using the graph-based feature extraction model, a Multilayer Perceptron classification algorithm has been implemented to classify 9000 Bangla text documents obtained from various web sources. The details of the experimental results are presented in Section 5.4.

### 5.3.1 Data Collection

It has been observed from the survey of the literature that the domain-based Bangla text documents are not available in a sufficient quantity, so we had to generate our own database for carrying out the text categorization task. Data being the most important element in the outcome of any experiment,

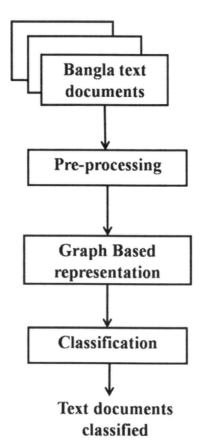

**FIGURE 5.1**
Pipeline of the proposed methodology.

**TABLE 5.1**

Text Document Distribution for Each Domain

| Domain | Number of Documents |
|---|---|
| Business | 1100 |
| Entertainment | 1100 |
| Food | 1000 |
| Medical | 1000 |
| State News | 1400 |
| Sports | 1400 |
| Science and Technology | 1000 |
| Travel | 1000 |
| **Total** | **9000** |

it should be gathered adequately both in terms of quality and quantity. Development of a database requires much attention so that errors do not come up while conducting the experiment and the appropriate results may be reached. The text documents involved in the experiment were acquired from various online Bangla news corpora: magazines and webpages that cover eight text domains, i.e. Business (B), Sports (SP), Entertainment (E), Science and Technology (ST), Medical (M), Travel (T), Food (F) and State News (P), as stated by leading Bangla newspapers. The distribution of Bangla text documents for each of the eight domains is provided in Table 5.1. The web sources of the text documents are anandabazar.com, bartamanpatrika. com, ebela.in, deshirecipe.wordpress.com, ebanglarecipe.com, recipebangla. com, banglaadda.com, amaderchhuti.com, ebanglatravel.com.

## 5.3.2 Pre-Processing

Linear sequences of representations such as characters, words and phrases are involved in the formation of digital text documents. For a given sequence of characters in a respective text document, tokenization needs to be carried out before the implementation of the feature extraction and selection approaches. Tokenization is the task of segregating the sentences into pieces, called tokens. Without chopping the sentences clearly into the tokens, further processing is not possible. Here, the splitting of sentences was performed based on a 'space' delimiter and the total number of words retrieved from the dataset was 2,806,159. The tokenized data enhances the processing of the system, makes it more efficient and reduces the burden on system resources. This proves to be an advantage in systems relying on better performances. Since not all the words extracted from the dataset are relevant or informative enough to be in the feature set, the removal of stopwords from this set is necessary to filter the data. In the present experiment, punctuations, pronouns, English equivalent words, postpositions, English and Bangla numerals, interjections, some adjectives, conjunctions, some adverbs and all the articles have been considered stopwords and the number of tokens acquired is thus 1,310,119. Consideration of stopwords depends on a specific language as well as a specific problem.

### 5.3.3 Graph-Based Representation

In the present experiment, a graph-based text representation scheme has been implemented based on a weighted graph algorithm. Graph models have been used in several fields [27]. Then a classification algorithm was used to assign the text category to the text documents in the dataset.

A graph Z is denoted by 3-tuple: Z = (V, E, W), where V is a set of nodes or vertices, E is a collection of weighted edges connecting the nodes and W represents the feature weight vector of edges.

- Node: Unique terms obtained from the training set using the feature extraction methods.
- Edge: Established depending on the occurrence of terms in text documents. If two terms arrive in a certain text document then an edge was assigned from the previous node to the later one.
- Feature Weight Vector: Determined using equation 1, where $W(p,q)$ represents the weight assigned to the term $p$ and $q$; $f(p,q)$ denotes the count of occurrence of the term $p$ and $q$ within a text document; $f(p)$ denotes the count of occurrence of the term $p$; and $f(q)$ denotes the count of occurrence of the term $q$ separately in that particular text document. The higher the weight $W(p,q)$, the stronger the edge between the two nodes is.

$$W(p,q) = \frac{f(p,q)}{f(p) + f(q) - f(p,q)} \tag{5.1}$$

The trend of the features for the eight distinct text classes is graphically depicted in Figure 5.2.

### 5.3.4 Classifier

A multilayer perceptron (MLP), one of the popular and commonly used classifiers in pattern recognition problems [28–31], is a feed-forward artificial neural network architecture that plots the input data to a set of output data. It involves at least three surfaces of points bounded with the number of neurons in every surface, which are presented as a directed graph that employs a non-linear activation function. Its multiple surfaces and non-linear activation function make MLP capable of classifying data that are not linearly separable, and also of differentiating them from a linear perceptron. In our experiment, we have opted for the configuration of the MLP as i-hl-o (i.e. i denotes the number of attributes, hl represents the number of hidden layers and o is the number of output text

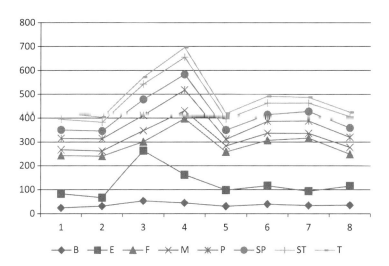

**FIGURE 5.2**
Feature trends of eight text categories.

categories). It applies a back propagation method to train the architecture. The number of neurons to be in the hidden layer is computed as a consequence of the expanse of the feature vector of the training dataset. Each neuron in this architecture has a sigmoid function and the architecture has been trained based on the number of iterations being 1,500, learning rate 0.3 and momentum 0.2.

## 5.4 Results and Analysis

The results obtained by carrying out the Bangla text classification task on 9,000 documents, from eight distinct categories based on K fold cross-validation as used in [32, 33], K being 5, are presented in this section. A maximum accuracy of 98.62% was obtained for 1,500 iterations on the reduced feature set by using the Multilayer Perceptron classification algorithm. In this experiment, the MLP was applied using WEKA [34], a widely used open source classification implementation. The confusion matrix obtained by applying MLP based on 1,500 iterations and five-fold cross-validation on the generated dataset is presented in Table 5.2. The rate of classification and misclassification has been achieved using the following equations.

$$\text{Classification} = \frac{\text{Correctly classified documents}}{\text{Total number of text documents}} * 100\% \qquad (5.2)$$

$$\text{Misclassification} = \frac{\text{Incorrectly classified documents}}{\text{Total number of text documents}} * 100 \qquad (5.3)$$

**TABLE 5.2**

Confusion Matrix Obtained Using MLP

|      | B    | E    | F   | M   | P    | SP   | ST  | T   |
|------|------|------|-----|-----|------|------|-----|-----|
| B    | 1065 | 1    | 0   | 3   | 13   | 1    | 14  | 3   |
| E    | 3    | 1088 | 0   | 0   | 4    | 4    | 1   | 0   |
| F    | 0    | 0    | 997 | 2   | 0    | 0    | 0   | 1   |
| M    | 1    | 1    | 2   | 986 | 8    | 0    | 2   | 0   |
| P    | 12   | 2    | 0   | 12  | 1370 | 2    | 1   | 1   |
| SP   | 1    | 4    | 0   | 0   | 3    | 1390 | 0   | 2   |
| ST   | 6    | 0    | 0   | 0   | 0    | 0    | 993 | 1   |
| T    | 3    | 1    | 1   | 1   | 6    | 0    | 1   | 987 |

Based on the confusion matrix, the average percentages of correctly classified text documents as well as the rate of misclassification for each domain are given in Table 5.3. The fall in the performance of the system based on the adopted methodology is due to the fact that there is a resemblance in the organization or the arrangement of the tokens in the text documents of two or more text categories, which leads to the misclassification of text documents. From Table 5.3, it can be observed that the text documents belonging to Business and State News domains have higher rates of misclassification than other text domains due to the fact that the contents of the texts of these domains are quite similar to each other.

The performance level achieved from the present experiment has also been compared with other commonly used classification algorithms, namely naive Bayes multinomial (NBMN), naive Bayes multivariate (NBMV), decision tree (J48) and K nearest neighbour (KNN) [35] to test the robustness of our system. The obtained accuracy for all the classifiers is provided in Table 5.4 and a comparative analysis based on precision, recall and $F_1$ measure is shown in Figure 5.3, from which it can be seen that the methodology adopted in the present experiment performs better using MLP in comparison with others in the case of Bangla text document classification. The worst was NBMV; KNN, J48 and NBMN obtained 2nd, 3rd, and 4th ranks respectively.

**TABLE 5.3**

Average Percentage of Correct and Incorrect Classifications

| Domains                | Correct Classification (%) | Incorrect Classification (%) |
|------------------------|----------------------------|------------------------------|
| Business               | 96.82                      | 3.18                         |
| Entertainment          | 98.90                      | 1.10                         |
| Food                   | 99.70                      | 0.30                         |
| Medical                | 98.60                      | 1.40                         |
| State News             | 97.85                      | 2.15                         |
| Sports                 | 99.28                      | 0.72                         |
| Science and Technology | 99.30                      | 0.70                         |
| Travel                 | 98.70                      | 1.30                         |
| **Average**            | **98.64**                  | **1.36**                     |

**TABLE 5.4**

Comparison of Classifiers

| Classifier | Accuracy (%) |
| --- | --- |
| MLP | **98.62** |
| NBMN | 96.49 |
| NBMV | 94.20 |
| J48 | 97.09 |
| KNN | 97.98 |

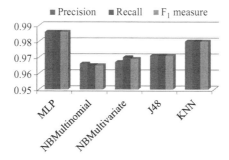

**FIGURE 5.3**
Comparison among classifiers.

## 5.4.1  Comparison with Existing Methods

The proposed work has been compared with some of the existing works in Bangla. In this comparison, only the schemes applied by other researchers to train their models in the existing works have been followed because of the unavailability of their databases. The experiments were evaluated using the MLP algorithm on our obtained database involving 9000 Bangla text documents. The obtained accuracy for all the methods is provided in Table 5.5 below. From the table, it can be seen that the approach proposed here performs well compared to all other techniques used for Bangla with respect to recognition accuracy. It can also be said that the performance of various

**TABLE 5.5**

Comparison of the Proposed Methodology with Existing Methods

| References | Feature Used | Size of Data Set | Accuracy (%) |
| --- | --- | --- | --- |
| Mansur et al. [21] | N-gram | 9000 | 93.79 |
| Mandal and Sen [22] | TF-IDF | 9000 | 96.21 |
| Kabir et al. [23] | TF-IDF | 9000 | 96.83 |
| Islam et al. [24] | TF-IDF & Chi Square | 9000 | 95.36 |
| **Proposed Method** | **Graph-Based** | **9000** | **98.62** |

distance metrics for text documents classification is quite encouraging compared to the traditional machine learning algorithms.

## 5.5 Conclusion

This chapter presents an approach that uses a multilayer perceptron classification algorithm based on a graph-based text representation scheme for classifying Bangla text documents into their respective text categories. The results were also compared with other commonly used classifiers and it has been observed from the experiments that the multilayer perceptron performs better compared to other algorithms for classifying Bangla text documents. In future, this experiment can be expanded to include a larger number of domains of Bangla texts as well as text categories. Also new hybrid approaches can be introduced for classifying the text documents to their predetermined text categories. We also plan to experiment with other machine learning techniques like those based on deep learning [36] and active learning [37] approaches to obtain better results. Also, a dynamic programming-based approach as stated in [38, 39] and ELM in [40] can be explored in text categorization tasks. In future, this study can be explored on the datasets of other languages as well.

## Acknowledgment

One of the authors thanks DST for its support in the form of an INSPIRE fellowship.

## References

1. Dasondi, V., Pathak, M., Rathore, N.P.S.: An Implementation of Graph Based Text Classification Technique for Social Media, Proc. of CDAN, p. 7, 2016.
2. Guru, D.S., Suhil, M., Raju, L.N., Kumar, N.V.: An Alternative Framework for Univariate Filter Based Feature Selection for Text Categorization, *Pattern Recognition Letters*, 103, pp. 23–31, 2018.
3. Asim, M.N., Wasim, M., Ali, M.S., Rehman, A.: Comparison of Feature Selection Methods in Text Classification on Highly Skewed Datasets, Proc. of INTELLECT, p. 8, 2017.

4. Kowsari, K., Heidarysafa, M., Brown, D.E., Meimandi, K.J., Barnes, L.E.: RMDL: Random Multimodel Deep Learning for Classification, Proc. of ICISDM, p. 11, 2018.
5. Feng, G., Li, S., Sun, T., Zhang, B.A.: A Probabilistic Model Derived Term Weighting Scheme for Text Classification, *Pattern Recognition Letters*, 110, pp. 23–29, 2018.
6. Wang, Z., Liu, Z.: Graph-Based KNN Text Classification, Proc. of FSKD, pp. 2363–2366, 2010.
7. Wei, Z., Miao, D., Chauchat, J., Zhao, R., Li, W.: N-Grams Based Feature Selection and Text Representation for Chinese Text Classification, *International Journal of Computational Intelligence Systems*, 2, pp. 365–374, 2009.
8. Wenliang, C., Xingzhi, C., Huizhen, W., Jingbo, Z., Tianshun, Y.: Automatic Word Clustering for Text Categorization Using Global Information, Proc. of AIRS, pp. 1–11, 2005.
9. Mikawa, K., Ishidat, T.,Goto, M.: A Proposal of Extended Cosine Measure for Distance Metric Learning in Text Classification, Proc. of ICSMC, pp. 1741–1746, 2011.
10. Parvin, H., Dahbashi, A., Parvin, S., Minaei, B.: Improving Persian Text Classification and Clustering Using Persian Thesaurus, Proc. of ICDCAI, pp. 493–500, 2012.
11. Al-Tahrawi, M.: Arabic Text Categorization Using Logistic Regression, *International Journal of Intelligent Systems and Applications*, 6(6), pp. 71–78, 2015.
12. Al-Radaideh, Q.A., Al-Khateeb, S.S.: An Associative Rule-Based Classifier for Arabic Medical Text, *International Journal of Knowledge Engineering and Data Mining*, 3, pp. 255–273, 2015.
13. Haralambous, Y., Elidrissi, Y., Lenca, P.: Arabic Language Text Classification Using Dependency Syntax-Based Feature Selection, Proc. of ICALP, p. 10, 2014.
14. Ali, A., Ijaz, M.: Urdu Text Classification, Proc. of ICFIT, pp. 1–7, 2009.
15. Gupta, N., Gupta, V.: Punjabi Text Classification Using Naive Bayes, Centroid and Hybrid Approach, Proc. of SEANLP, pp. 109–122, 2012.
16. ArunaDevi, K., Saveeth, R.: A Novel Approach on Tamil Text Classification Using C-Feature, *International Journal of Scientific Research & Development*, 2, pp. 343–345, 2014.
17. Patil, J.J., Bogiri, N.: Automatic Text Categorization Marathi Documents, *International Journal of Advanced Research in Computer Science and Management Studies*, 3, pp. 280–287, 2015.
18. Patil, M., Game, P.: Comparison of Marathi Text Classifiers, ACEEE, *International Journal on Information Technology*, 4, pp. 11–22, 2014.
19. Sarmah, J., Saharia, N., Shikhar, K.: A Novel Approach for Document Classification Using Assamese Wordnet, Proc. of IGWC, pp. 324–329, 2012.
20. Mohanty, S., Santi, P., Mishra, R., Mohapatra, R., Swain, S.: Semantic Based Text Classification Using Wordnets: Indian Language Perspective, Proc. of ICEFCE, pp. 321–324, 2006.
21. Mansur, M., UzZaman, N., Khan, M.: Analysis of N-Gram Based Text Categorization for Bangla in a Newspaper Corpus, Proc. of ICCIT, p. 8, 2006.
22. Mandal, A.K., Sen, R.: Supervised Learning Methods for Bangla Web Document Categorization, *International Journal of Artificial Intelligence & Applications*, 5(5), pp. 93–105, 2014.

23. Kabir, F., Siddique, S., Kotwal, M.R.A., Huda, M.N.: Bangla Text Document Categorization Using Stochastic Gradient Descent (SGD) Classifier, Proc. of ICCCIP, pp. 1–4, 2015.

24. Islam, Md.S., Jubayer, F.E.Md., Ahmed, S.I.: A Support Vector Machine Mixed with TF-IDF Algorithm to Categorize Bengali Document, Proc. of ICECCE, pp. 191–196, 2017.

25. Dhar, A., Dash, N.S., Roy, K.: Classification of Text Documents through Distance Measurement: An Experiment with Multi-Domain Bangla Text Documents, Proc. of ICACCA, pp. 1–6, 2017.

26. Dhar, A., Dash, N.S., Roy, K.: Application of TF-IDF Feature for Categorizing Documents of Online Bangla Web Text Corpus, Proc. of FICTA, pp. 51–59, 2017.

27. Santosh, K.C.: g-DICE: Graph Mining-Based Document Information Content Exploitation, *International Journal on Document Analysis and Recognition*, 18(4), pp. 337–355, 2015.

28. Pembe, F.C., Gungor, T.: A Tree-Based Learning Approach for Document Structure Analysis and Its Application to Web Search, *Natural Language Engineering*, 2, pp. 569–605, 2014.

29. Obaidullah, Sk.M., Halder, C., Santosh, K.C., Das, N., Roy, K.: PHDIndic_11: Page-Level Handwritten Document Image Dataset of 11 Official Indic Scripts for Script Identification, *Multimedia Tools and Applications*, 77, pp. 1643–1678, 2017.

30. Obaidullah, S.M., Goswami, C., Santosh, K.C., Roy, K.: Separating Indic Scripts with `Matra' for Effective Handwritten Script Identification in Multi-Script Documents, *International Journal of Artificial Intelligence & Pattern Recognition (IJPRAI), World Scientific*, 31, p. 1753003 (17 pages), 2017.

31. Obaidullah, S.M., Santosh, K.C., Halder, C., Das, N., Roy, K.: Word-Level Multi-Script Indic Document Image Dataset and Baseline Results on Script Identification, *International Journal of Computer Vision and Image Processing (IJCVIP), IGI Global*, 7, pp. 81–94, 2017.

32. Obaidullah, S.M., Santosh, K.C., Halder, C., Das, N., Roy, K.: Automatic Indic Script Identification from Handwritten Documents: Page, Block, Line and Word-Level Approach, *International Journal of Machine Learning and Cybernetics*, pp. 1–20, 2017.

33. Santosh, K.C., Wendling, L.: Character Recognition Based on Non-Linear Multi-Projection Profiles Measure, *Frontiers of Computer Science*, 09(5), pp. 678–690, 2015.

34. Hall, M., Frank, E., Holmes, G., Pfahringer, B., Reutemann, P., Witten, I.H.: The Weka Data Mining Software: An Update, *ACM SIGKDD Explorations Newsletter*, 11(1), pp. 10–18, 2009.

35. Vajda, S., Santosh, K.C.: A Fast k-Nearest Neighbor Classifier Using Unsupervised Clustering, Proc. of Rtip2r, pp. 185–193, 2016.

36. Ukil, S., Ghosh, S., Obaidullah, S.M., Santosh, K.C., Roy, K., Das, N., *Deep Learning for Word-Level Handwritten Indic Script Identification*. arXiv Preprint arXiv:1801.01627, 2018.

37. Bouguelia, M.R., Nowaczyk, S., Santosh, K.C., Verikas, A.: Agreeing to Disagree: Active Learning with Noisy Labels without Crowdsourcing, *International Journal of Machine Learning and Cybernetics*, 9, p. 13, 2017.

38. Santosh, K.C.: Character Recognition Based on dtw-Radon, Proc. of ICDAR, pp. 264–268, 2011.

39. Santosh, K.C., Nattee, C., Lamiroy, B.: Spatial Similarity Based Stroke Number and Order Free Clustering, Proc. of ICFHR, pp. 652–657, 2010.
40. Obaidullah, S.M., Bose, A., Mukherjee, M., Santosh, K.C., Das, N., Roy, K.: Extreme Learning Machine for Handwritten Indic Script Identification in Multiscript Documents, *Journal of Electronic Imaging*, 27(5), p. 51214, 2018.

# 6

# A Study of Distance Metrics in Document Classification

Ankita Dhar, Niladri Sekhar Dash and Kaushik Roy

## CONTENTS

## 6.1 Introduction

Computers are able to perform an efficient and automated job using various approaches of data mining and machine learning. The volume of digital text data has suddenly increased with the rapid growth of the Internet in recent years. In addition, the manual investigation of each text data is a laborious, error-prone and time consuming task. Thus, there is an urgent need to deliver an efficient automatic text categorization system for these large amounts of unstructured text resources. Text categorization has become a major task in text mining problems. Several text categorization techniques have been developed in recent years in various languages, but fail to meet the requirements for Indian languages, especially for Bangla. Therefore this paper aims to present a text categorization technique that performs the classification task for Bangla text documents. The proposed methodology is a text categorization problem that takes the pre-labeled Bangla text documents obtained from various news corpora, online magazines and webpages as the input; these documents are the training data required to train the model classifier based on the various distance measurement algorithms and classify them into the respective text categories that are considered for this work.

In this chapter, some of the distance measurement algorithms such as squared Euclidean distance, Manhattan distance, Mahalanobis distance, Minkowski distance, Chebyshev distance and Canberra distance have been explored for Bangla text categorization for the first time, their applicability in this task examined using a dataset of 9000 Bangla text documents generated from various web news corpora, magazines and webpages covering eight text categories. Also, some standard and widely used distance measurement algorithms were studied, namely Cosine Similarity and Euclidean distance on the datasets, for the purposes of comparison.

The rest of the paper has the following structure. The related work is discussed in Section 6.2, followed by the proposed methodology in Section 6.3, which includes discussion on data collection, pre-processing techniques, feature extraction and selection approaches and various distance measurement algorithms. In Section 6.4, the results are discussed along with a comparative analysis among various distance measurement algorithms as well as the comparison of the proposed approach with existing methods. In Section 6.5, the paper is concluded showing some future directions in this field.

## 6.2 Literature Survey

### 6.2.1 Indo–European

English text classification has gained immense acceptance among researchers all over the place. In the Indo–European language family, besides English, a few works have been carried out on Persian and Urdu as well. For instance, Lin et al. [1] proposed a new similarity measure named similarity measure for text processing (SMTP) by considering three cases: (i) the feature appearing in both text documents, (ii) the feature appearing in only one text document and (iii) the feature appearing in none of the text documents. They carried out their experiment on WebKB, Reuters-8 and RCV1 based on TF-IDF, and obtained accuracies of 0.8467, 0.9488 and 0.7130 for the value of K = 11, 11 and 8, respectively. In another experiment, Wandabwa et al. [2] extracted the most relevant features by assigning them weights in respect to their contents. Then they estimated the distance between the features with high ranks in the dataset for the classification task. They experimented on a dataset comprised of 15,000 documents of random Wikipedia personalities having elements like URI, name and text. An experiment by Wu et al. [3] presented a method of balancing between over-weighting and under-weighting the terms for text classification using 20 Newsgroups, RT-2K, IMDB and Reuters-21578 datasets. Koswari et al. [4] proposed a technique called random multimodel deep learning (RMDL) that solves the issue of perceiving the ideal deep learning network by enhancing accuracy through ensembles of deep learning networks. They experimented using the IMDB dataset and obtained a maximum accuracy of 90.79% using a 15 RDLs model, and also observed that RDML performs way better in comparison to the traditional approaches. Feng et al. [5] presented a new scheme for weighting the terms based on a probabilistic model using a matching score function, and observed the effectiveness of the methods using K-NN and SVM on small and large texts, respectively. In the case of Persian, Parvin et al. [6] classified Persian text documents from five text categories by using normalized mutual information along with a thesaurus and obtained a maximum accuracy of 81.45% for the value K = 1. Ali and Ijaz [7] provided a comparative study among the statistical methods for classifying Urdu text documents using NB and SVM. Mainly they have preprocessed the documents using language-specific techniques for developing a standard and less dimensional feature vector. From the experiments, it can be seen that the performance increases when the value of n increases up to 3, but decreases when the values reaches 4.

### 6.2.2 Sino–Tibetan

Besides the cases of English, Persian and Urdu, various works carried out for Chinese can be referred to here. To name a few, Wei et al. [8] conducted experiments in a Chinese corpus consisting of 14,150 Chinese text documents from 12 classes based on three parameters: (i) comparison of the performance among absolute and relative text frequency and N-gram frequency; (ii) comparison between sparseness and feature correlation; and (iii) performance comparison among 0/1 logical value, N-gram frequency numeric value and TF-IDF value. Wenliang et al. [9] used the NEU TC dataset of 14,459 texts from 37 domains and applied document frequency and chi square statistics to develop the class distribution of tokens. They also used a global clustering model, referred to as a global CM, for Chinese text categorization.

### 6.2.3 Japonic

Mikawa et al. [10] proposed an efficient approach using extended cosine measure for classification of the Mainichi newspaper and customer reviews in Japanese. The approach measures the distance between two documents based on a vector space model.

### 6.2.4 Afro–Asiatic

Various techniques of text categorization have also been explored for Arabic. To mention a few, Boukil et al. [11] used convolutional neural networks for categorizing Arabic texts after extracting features using TF-IDF scheme. Al-Tahrawi and Al-Khatib [12] worked with polynomial neural networks on the Al-Jazeera dataset and achieved encouraging results. Raho et al. [13] studied the performances of various learning models for different cases on the BBC Arabic dataset and presented the results which are quite useful.

### 6.2.5 Dravidian

Recent research on text classification in Dravidian languages is briefly discussed here. For the classification of Tamil text documents, ArunaDevi and Saveeth [14] used the C-feature extraction technique, which takes pairs of terms into consideration.

### 6.2.6 Indo–Aryan

Recent research carried out for text classification in Indo–Aryan languages is briefly discussed here. For instance, Gupta and Gupta [15] combined NB- and Ontology-based techniques to classify 184 Punjabi text documents extracted from various news corpora covering seven sub-domains of sports. Their experiment has shown that the combined approach performs better compared to other approaches. Patil and Bogiri [16] performed an automatic Marathi text document categorization on 200 documents of 20 categories based on user profiles using the label induction grouping (LINGO) algorithm. Rakholia and Saini [17] implemented a naive Bayes model for classifying 280 Gujarati documents from six categories for two cases: with and without feature selection methods and obtained accuracies of 88.96% and 75.74% respectively. Classification of texts using language-specific WordNets for Assamese and Sanskrit was performed by Sarmah et al. [18] and Mohanty et al. [19]. In the case of Assamese [18], four classes were considered and an encouraging result of 90.27% accuracy was obtained for the same. However, the state of text classification approaches tells a sorry tale for Bangla. The lack of information for this field shows that it will still be a long time before we notice the urgent need for more research in this language. Some of the work performed on this language are reported here. Alam and Islam [20] studied text based features on 376,226 articles and obtained precision of 0.95 using a logistic regression model. Mandal and Sen [21] experimented with four supervised learning methods to classify 1000 labeled documents with a total of 22,218 tokens in five categories. They achieved maximum accuracy of 89.14% with the SVM method. On the other hand, a Stochastic Gradient Descent (SGD) classifier was used by Kabir et al. [22] to classify documents from nine distinct domains and achieved an accuracy of 93.85%, which is better compared to the previous one. In another work [23], an accuracy of 92.57% was observed based on a TF-IDF and SVM algorithm for classifying Bangla text documents of 12 text domains. Dhar et al. [24] worked with text representation schemes along with a multilayer perceptron model on 4000 articles and achieved 98.03% accuracy. In another work [25], the author explored the optimization problem on 9000 articles by modifying the Cuckoo search algorithm with weighting schemes and obtained an accuracy of 98%. The brief survey presented here certainly shows that we still need to go a long way to develop an automatic text categorization system for Bangla language users, in order to serve their purposes efficiently.

## 6.3 Proposed Methodology

An overview of the proposed methodology is illustrated in Figure 6.1. On a feature set acquired using the feature extraction approach (term frequency-inverse document frequency-inverse class frequency), various distance measures have been implemented to classify 9000 documents obtained from various web sources in Bangla. The details of the experimental results are presented in Section 6.4.

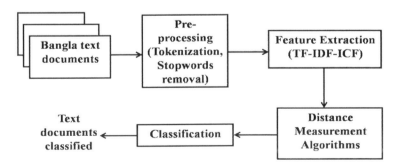

**FIGURE 6.1**
Bird's eye view of the proposed method.

## 6.3.1 Data Collection

Data both in terms of quality and quantity acts as an important factor in the outcome of any experiment. Development of a database requires much attention so that errors do not come up while executing the experiment and the appropriate outcome may be attained. However, from the brief literature survey, it has been observed that the availability of category-based Bangla text documents is not adequate, so we had to develop our own database for carrying out the task. The text documents involved in the experiment were acquired from various online Bangla news corpora, magazines and webpages, covering eight text domains: Business (B), Sports (SP), State News (P), Entertainment (E), Travel (T), Medical (M), Food (F) and Science and Technology (ST), as stated by leading newspapers. The distribution of text documents for each of the eight domains are provided in Table 6.1. The web sources of the text documents are anandabazar.com, bartamanpatrika.com, ebela.in, deshirecipe.wordpress.com, ebanglarecipe.com, recipebangla.com, banglaadda.com, amaderchhuti.com, ebanglatravel.com.

## 6.3.2 Pre-Processing

Digital text documents are formed of a linear series of symbols such as characters, words and phrases. For a given sequence of characters in its respective text document, before implementing the feature extraction and selection approaches, tokenization needs to be performed. Tokenization is the process of chopping the sentences into pieces, called tokens. Without segmenting the sentences clearly into these units, further processing is not possible. Conceptually, the sentences were split based on a 'space' delimiter. After tokenization, the count of terms after retrieval is 2,806,159. The tokenized data allows the processing of the system more quickly and efficiently and also reduces pressure on system resources. This proves to be an advantage in systems relying on better performance. Since not all the tokens extracted

**TABLE 6.1**

Text Document Distribution for Each Domain

| Domain | Number of Documents |
| --- | --- |
| Business | 1100 |
| Entertainment | 1100 |
| Food and Recipe | 1000 |
| Medical | 1000 |
| State News | 1400 |
| Sports | 1400 |
| Science and Technology | 1000 |
| Travel | 1000 |
| **Total** | **9000** |

from the dataset are relevant or informative enough to be preserved in the feature set, stopword removal is necessary to clean and filter the data. Here, in the present experiment, punctuations, pronouns, English equivalent words, postpositions, English and Bangla numerals, interjections, some adjectives, conjunctions, some adverbs and all the articles have been considered stopwords, resulting in 1,310,119 tokens acquired. The subject of considering stopwords is language specific as well as problem-specific.

### 6.3.3 Feature Extraction and Selection

The standard and widely used weighting scheme, term frequency–inverse document frequency (TF-IDF), is modified into a new approach and named TF-IDF-ICF, adding an inverse class frequency (ICF) into it to develop the feature vector from the dataset [26]. An inverse document frequency (IDF) is basically used for evaluating the measure of differentiation of a term among relevant and non-relevant text documents. The experiments performed earlier have shown that there is more likelihood of a token occurring in text documents of domain $x$ than in documents that do not belongs to $x$ in cases where the term is a target term of $x$. Depending on this observation, ICF is proposed to measure the discerning power of a term to a domain.

The TF counts the presence of a token $t$ in a specific text $d$. Here, log normalization is considered to become the normalized term frequency for each term $t$.

$$\text{TF} = \log \frac{O_{t,d}}{\sum_{t \in d} O_{t,d}} \tag{6.1}$$

The IDF generally predicts the relevancy of a term $t$ in the total dataset. The document frequency (DF) counts the presence of a term $t$ in a number of texts in the dataset using the equation below where $N$ represents the dataset.

$$\text{IDF} = \log \frac{N}{\text{DF}(t)} \tag{6.2}$$

The class frequency (CF) counts the presence of a token $t$ in the total statistics of the domains. ICF is estimated based on the following equation where $C$ is the overall categories being considered in the experiment.

$$\text{ICF} = \log \frac{N}{\text{CF}(t)} \tag{6.3}$$

The proposed TF-IDF-ICF is a novel scheme that considers the presence of a token $t$ in a domain $c$ with TF-IDF in consideration, while measuring the weight for $t$ using Equation 6.4. In every case, normalized values for the terms were considered to get a more specific and relevant feature vector.

$$\text{TF-IDF-ICF} = \text{TF} * \text{IDF} * \text{ICF} \tag{6.4}$$

The trend of the features for the eight distinct text classes considered for the present experiment is graphically illustrated in Figure 6.2.

### 6.3.4 Distance Measurement

Distance measures are an important affair in many machine learning tasks. The purpose of the distance measurement is to estimate the similarity or dissimilarity between two terms. The distance measures considered are squared Euclidean distance, Manhattan distance, Mahalanobis distance, Minkowski distance, Chebyshev distance and Canberra distance.

#### 6.3.4.1 Squared Euclidean Distance

The squared Euclidean distance measure represents the same formula as the Euclidean distance measure, but the difference is that it is squared in order to assign gradually increasing weight on terms that are at longer distances. It is estimated using the equation below.

$$\text{Dis}^2(i,j) = (i_1 - j_1)^2 + (i_2 - j_2)^2 + \ldots + (i_n - j_n)^2 \tag{6.5}$$

The squared Euclidean distance considers the sum of the squared differences between two terms, corresponding to a text document.

#### 6.3.4.2 Manhattan Distance

The Manhattan distance has various names like city block distance, L1 distance or absolute value distance. It measures the distance between two terms along the axes at right angles and is calculated using Equation 6.6. This distance measure is less influenced by the deviation, compared to the Euclidean and squared Euclidean measures.

$$\text{Dis}(i,j) = \|i - j\| = \sum_{l=1}^{n} |i_l - j_l| \tag{6.6}$$

#### 6.3.4.3 Mahalanobis Distance

The Mahalanobis distance takes the mean or centroid of the varied data into consideration. It measures the relative distance between two terms with respect to the centroid. The Mahalanobis distance is larger for terms which are farther apart from the centroid. The distance is represented by Equation 6.7 where SD is the standard deviation of the $i_l$ and $j_l$ over the dataset. The

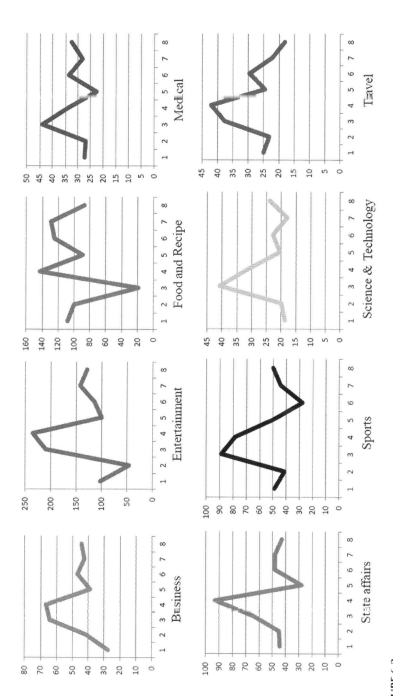

**FIGURE 6.2**
Feature trends of eight text categories.

covariance matrix presents the covariance from being identified with the objects. The reason behind the use of covariance is to get the consequences of two or more objects together. It is generally used in classification and clustering tasks that require the development of correlation between various clusters of objects.

$$\text{Dis}(i, j) = \sqrt{\sum_{l=1}^{n} \frac{(i_l - j_l)^2}{SD_l^2}} \qquad (6.7)$$

### 6.3.4.4 Minkowski Distance

The Minkowski distance can be recognized as a generalization of both the Euclidean distance and the Manhattan distance computed using the following equation where m is a positive integer value considered to be 1 or 2.

$$\text{Dis}(i, j) = \left( \sum_{l=1}^{n} |i_l - j_l|^m \right)^{1/m} \qquad (6.8)$$

### 6.3.4.5 Chebyshev Distance

The Chebyshev distance can be defined on the feature vectors to measure the greatest of the differences between two vector spaces along the standard coordinates. It is measured using the equation below.

$$\text{Dis}(i, j) = \max_l |i_l - j_l| \qquad (6.9)$$

### 6.3.4.6 Canberra Distance

The Canberra distance explores the sum of a series of a fraction difference between the coordinates of a pair of variables. Each variable has a value between 0 and 1 which is very susceptible to a small change. The Canberra distance for the given vectors *i* and *j* is represented using Equation 6.10. It is a numerical measure of the distance between pairs of terms in a vector space.

$$\text{Dis}(i, j) = \sum_{l=1}^{n} \frac{|i_l - j_l|}{|i_l| + |j_l|} \qquad (6.10)$$

## 6.4 Results and Discussion

The proposed methodology has been implemented on the dataset acquired, which is comprised of 9000 documents in Bangla from eight distinct classes

**TABLE 6.2**

Confusion Matrix Obtained Using Mahalanobis
Distance

|      | B    | E    | F   | M   | P    | SP   | ST  | T   |
|------|------|------|-----|-----|------|------|-----|-----|
| B    | 1074 | 0    | 0   | 6   | 14   | 0    | 6   | 0   |
| E    | 2    | 1090 | 0   | 0   | 4    | 2    | 0   | 2   |
| F    | 0    | 0    | 996 | 0   | 0    | 0    | 0   | 4   |
| M    | 4    | 0    | 0   | 981 | 5    | 5    | 5   | 0   |
| P    | 11   | 6    | 0   | 9   | 1361 | 8    | 5   | 0   |
| SP   | 8    | 6    | 0   | 2   | 8    | 1372 | 0   | 4   |
| ST   | 4    | 0    | 0   | 6   | 0    | 4    | 986 | 0   |
| T    | 0    | 0    | 5   | 0   | 0    | 2    | 0   | 993 |

(Business, Entertainment, Food, Medical, State News, Sports, Science and Technology and Travel) based on a TF-IDF-ICF feature extraction scheme using various distance measurement algorithms (such as squared Euclidean distance, Manhattan distance, Mahalanobis distance, Minkowski distance, Chebyshev distance and Canberra distance). K fold cross-validation was used in [27, 28], K being 5. From the experiments, it can be seen that the Mahalanobis distance outperforms all other distance metrics; the confusion matrix using this distance measure is provided in Table 6.2. The domain-wise accuracy based on various parameters obtained after performing classification is depicted through Figure 6.3.

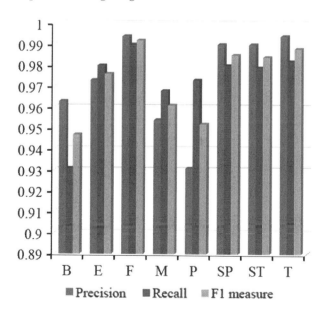

**FIGURE 6.3**
Domain-wise evaluation based on various parameters.

**TABLE 6.3**

Average Percentage of Correct and Incorrect Classifications

| Domains | Correct Classification (%) | Incorrect Classification (%) |
|---|---|---|
| Business | 97.64 | 2.36 |
| Entertainment | 99.10 | 0.90 |
| Food | 99.60 | 0.40 |
| Medical | 98.10 | 1.90 |
| State News | 97.21 | 2.79 |
| Sports | 98.00 | 2.00 |
| Science and Technology | 98.60 | 1.40 |
| Travel | 99.30 | 0.70 |
| **Average** | **98.44** | **1.56** |

The reason behind the fall in accuracy in the adopted methodology is that in some cases the measure of similarity among the tokens in the text documents of two domains is much more, which cannot be neglected or ignored and which leads to the misclassification of documents. Based on the confusion matrix obtained using the Mahalanobis distance, the rate of correctly classified text documents as well as the rate of misclassification for each domain is given in Table 6.3, calculated using the following equations.

$$\text{Classification} = \frac{\text{Correctly classified documents}}{\text{Total number of text documents}} * 100\% \qquad (6.11)$$

$$\text{Misclassification} = \frac{\text{Incorrectly classified docuemnts}}{\text{Total number of text docuemnts}} * 100 \qquad (6.12)$$

The comparative analysis of all the distance measurement algorithms considered for this experiment is presented in terms of accuracy in Table 6.4. It can be seen that the Mahalanobis distance performs best, while the Chebyshev distance was the worst; the Minkowski distance, Canberra distance, squared

**TABLE 6.4**

Performance of Different Distance Metrics

| Classification Algorithms | Accuracy (%) |
|---|---|
| Mahalanobis Distance | 98.37 |
| Minkowski Distance | 98.00 |
| Canberra Distance | 97.81 |
| Squared Euclidean Distance | 97.52 |
| Manhattan Distance | 97.11 |
| Chebyshev Distance | 96.26 |

**TABLE 6.5**

Comparison of the Proposed Methodology to Existing Methods

| References | Feature Used | Size of Data Set | Accuracy (%) |
|---|---|---|---|
| Mansur et al. [20] | N-gram | 9000 | 94.21 |
| Mandal and Sen [21] | TF-IDF | 9000 | 95.69 |
| Kabir et al. [22] | TF-IDF | 9000 | 96.47 |
| Islam et al. [23] | TF-IDF and Chi Square | 9000 | 95.84 |
| **Proposed Method** | **TF-IDF-ICF** | **9000** | **98.37** |

Euclidean distance and Manhattan distance obtained 2nd, 3rd, 4th and 5th ranks, respectively.

## 6.4.1 Comparison with Existing Methods

The proposed work has been compared to some of the existing works in Bangla. In this comparison, only the schemes applied by other researchers to train their model in the existing works have been followed because of the unavailability of their databases. The experiments were evaluated using the Mahalanobis distance measurement algorithm on our database of 9000 Bangla text documents. The obtained accuracy for all the methods is provided in Table 6.5 below. From the table, it can be seen that the approach proposed here outperforms all other existing methods for Bangla in terms of recognition accuracy. It can also be said that the performance of various distance metrics for text document classification is quite encouraging compared to the traditional machine learning algorithms.

## 6.5 Conclusion

This chapter presents an approach that uses various distance metrics (namely squared Euclidean, Manhattan, Mahalanobis, Minkowski, Chebyshev and Canberra distance measures) based on a feature extraction scheme (TF-IDF-ICF) to classify Bangla text documents into their respective text categories. It can be observed from the experiments that the Mahalanobis distance measure performs better compared to other metrics for classifying text documents in Bangla. In future, this experiment can be expanded to include more Bangla text datasets as well as more domains or text categories. Also, new hybrid approaches can be introduced to classify text documents into their predetermined text categories. We also plan to experiment with other machine learning techniques like those based on deep learning [29] and active learning [30] approaches to obtain better results. A Dynamic programming-based approach, as stated in [31, 32] and ELM in [33], can also be explored in text categorization. In future, this study can be explored on the datasets of other languages as well.

## Acknowledgment

One of the authors thanks DST for its support in the form of an INSPIRE fellowship.

## References

1. Lin, Y.S., Jiang, J.Y., Lee, S.J.: A Similarity Measure for Text Classification and Clustering, *IEEE Transactions on Knowledge and Data Engineering*, 26(7), pp. 1575–1590, 2014.

2. Wandabwa, H., Zhang, D., Sammy, K.: Text Categorization Via Attribute Distance Weighted k-Nearest Neighbor Classification, Proc. of ICIT, pp. 225–228, 2016.

3. Wu, H., Gu, X., Gu, Y.: Balancing Between Over-Weighting and Under-Weighting in Supervised Term Weighting, *Information Processing and Management*, 53(2), pp. 547–557, 2017.

4. Kowsari, K., Heidarysafa, M., Brown, D.E., Meimandi, K.J., Barnes, L.E.: RMDL: Random Multimodel Deep Learning for Classification, Proc. of ICISDM, p. 11, 2018.

5. Feng, G., Li, S., Sun, T., Zhang, B.A.: A Probabilistic Model Derived Term Weighting Scheme for Text Classification, *Pattern Recognition Letters*, 110, pp. 23–29, 2018.

6. Parvin, H., Dahbashi, A., Parvin, S., Minaei, B.: Improving Persian Text Classification and Clustering Using Persian Thesaurus, Proc. of ICDCAI, pp. 493–500, 2012.

7. Ali, A., Ijaz, M.: Urdu Text Classification, Proc. of ICFIT, pp. 1–7, 2009.

8. Wei, Z., Miao, D., Chauchat, J., Zhao, R., Li, W.: N-Grams Based Feature Selection and Text Representation for Chinese Text Classification, *International Journal of Computational Intelligence Systems*, 2, pp. 365–374, 2009.

9. Wenliang, C., Xingzhi, C., Huizhen, W., Jingbo, Z., Tianshun, Y.: Automatic Word Clustering for Text Categorization Using Global Information, Proc. of AIRS, pp. 1–11, 2005.

10. Mikawa, K., Ishidat, T., Goto, M.: A Proposal of Extended Cosine Measure for Distance Metric Learning in Text Classification, Proc of ICSMC, pp. 1741–1746, 2011.

11. Boukil, S., Biniz, M., El Adnani, F., Cherrat, L., El Moutaouakkil, A.E.: Arabic Text Classification Using Deep Learning Technics. *International Journal of Grid and Distributed Computing*, 11, pp.103–114, 2018.

12. Al-Tahrawi, M.M., Al-Khatib, S.N.: Arabic Text Classification Using Polynomial Networks. *Journal of King Saud University-Computer and Information Sciences*, 27, pp. 437–449, 2015.

13. Raho, G., Al-Shalabi, R., Kanaan, G., Nassar, A.: Different Classification Algorithms Based on Arabic Text Classification: Feature Selection Comparative Study. *International Journal of Advanced Computer Science and Applications Ijacsa*, 6, pp. 23–28, 2015.

14. ArunaDevi, K., Saveeth, R.: A Novel Approach on Tamil Text Classification Using C-Feature, *International Journal of Scientific Research & Development*, 2, pp. 343–345, 2014.

15. Gupta, N., Gupta, V.: Punjabi Text Classification Using Naive Bayes, Centroid and Hybrid Approach, Proc. of SEANLP, pp. 109–122, 2012.

16. Patil, J.J., Bogiri, N.: Automatic Text Categorization Marathi Documents, *International Journal of Advanced Research in Computer Science and Management Studies*, 3, pp. 280–287, 2015.

17. Rakholia, R.M., Saini, J.R.: Classification of Gujarati Documents Using Naïve Bayes Classifier. *Indian Journal of Science and Technology*, 5, pp. 1–9, 2017.

18. Sarmah, J., Saharia, N., Shikhar, K.: A Novel Approach for Document Classification Using Assamese Wordnet, Proc. of IGWC, pp. 324–329, 2012.

19. Mohanty, S., Santi, P., Mishra, R., Mohapatra, R., Swain, S.: Semantic Based Text Classification Using Wordnets: Indian Language Perspective, Proc. of ICECCE, pp. 321–324, 2006.

20. Mansur, M., UzZaman, N., Khan, M.: Analysis of N-Gram Based Text Categorization for Bangla in a Newspaper Corpus, Proc. of ICCIT, p. 8, 2006.

21. Mandal, A.K., Sen, R.: Supervised Learning Methods for Bangla Web Document Categorization, *International Journal of Artificial Intelligence & Applications*, 5(5), pp. 93–105, 2014.

22. Kabir, F., Siddique, S., Kotwal, M.R.A., Huda, M.N.: Bangla Text Document Categorization Using Stochastic Gradient Descent (SGD) Classifier, Proc. of ICCCIP, pp. 1–4, 2015.

23. Islam, Md.S., Jubayer, F.E.Md., Ahmed, S.I.: A Support Vector Machine Mixed with TF-IDF Algorithm to Categorize Bengali Document, Proc. of ICECCE, pp. 191–196, 2017.

24. Dhar, A., Dash, N.S., Roy, K.: Categorization of Bangla Web Text Documents Based on TF-IDF-ICF Text Analysis Scheme, Proc. of CSI, pp. 477–484, 2018.

25. Dhar, A., Dash, N.S., Roy, K.: Efficient Feature Selection Based on Modified Cuckoo Search Optimization Problem for Classifying Web Text Documents, Proc. of RTIP2R, pp. 640–651, 2019.

26. Dhar, A., Dash, N.S., Roy, K.: Classification of Bangla Text Documents Based on Inverse Class Frequency, Proc. of IoT-SIU, pp. 1–6, 2018.

27. Obaidullah, S.M., Santosh, K.C., Halder, C., Das, N., Roy, K.: Automatic Indic Script Identification from Handwritten Documents: Page, Block, Line and Word-Level Approach, *International Journal of Machine Learning and Cybernetics*, pp. 1–20, 2017.

28. Santosh, K.C., Wendling, L.: Character Recognition Based on Non-Linear Multi-Projection Profiles Measure, *Frontiers of Computer Science*, 9(5), pp. 678–690, 2015.

29. Ukil, S., Ghosh, S., Obaidullah, S.M., Santosh, K.C., Roy, K., Das, N.: *Deep Learning for Word-Level Handwritten Indic Script Identification*. arXiv Preprint arXiv:1801.01627, 2018.

30. Bouguelia, M.R., Nowaczyk, S., Santosh, K.C., Verikas, A.: Agreeing to Disagree: Active Learning with Noisy Labels without Crowdsourcing, *International Journal of Machine Learning and Cybernetics*, 9, p. 13, 2017.
31. Santosh, K.C.: Character Recognition Based on dtw-Radon, Proc. of ICDAR, pp. 264–268, 2011.
32. Santosh, K.C., Nattee, C., Lamiroy, B.: Spatial Similarity Based Stroke Number and Order Free Clustering, Proc. of ICFHR, pp. 652–657, 2010.
33. Obaidullah, S.M., Bose, A., Mukherjee, M., Santosh, K.C., Das, N., Roy, K.: Extreme Learning Machine for Handwritten Indic Script Identification in Multiscript Documents, *Journal of Electronic Imaging*, 27(5), p. 51214, 2018.

# 7

# A Study of Proximity of Domains for Text Categorization

Ankita Dhar, Niladri Sekhar Dash and Kaushik Roy

## CONTENTS

## 7.1 Introduction

In the last few years, text document management systems based on contenthave gained tremendous attention in the field of computer and information science. The reasons behind this demand are the availability of digital text documents at a huge scale and the need to access these documents in more efficient manner. Thus, the emergenceof 'text categorization' (TC), which can also be referred to as 'text classification' or 'topic spotting'. Text categorization is a dynamic research domain of text mining that refers to the task of assigning text documents to their respective categories using some classification techniques for efficiently managing information. If the text document is categorized, then searching for and retrieving information from these texts will be quick and effective. The prime goal of text categorization is to classify a random text document to its category. The text categorization can be either single-label or multi-label: in the former case the text document will be classified with only one class, whereas in the latter it will fit into more than one category.

There are now various applications of TC in numerous frames. Some of the applications are document indexing depending on vocabulary, document filtering, word sense disambiguation and hierarchical classification of online resources. Normally, in recent scenarios, TC serves multiple purposes that require flexible access to and selection, organization and modification of the documents. It is also applicable where web automation is associated with the search and retrieval of data in the mode of text.

In the 1990s, machine learning approaches to text classification gained popularity. TC has a deep past. Several techniques have been explored in this field to enhance the performance in several languages, but very few approaches have been proposed for Indian languages, especially for Bangla. The common and widely used text representation is a bag-of-words approach that sometimes uses phrases, sequences of words or N-grams. Mostly, while working with texts, the prime focus is on words or N-grams that need to be extracted from the dataset before the application of feature selection or feature weighting schemes. Instead of classifying the text documents manually or being based on handcrafted classification rules, text categorization uses machine learning techniques which are trained by classification rules based on labeled training documents.

Here, in the present experiment, the unique bag of words (UBoWs) and the proximity between the text categories have been considered in order to increase the efficiency of a system for classifying Bangla text documents automatically. The proximity of categories is determined based on the UBoWs model. Then, a scoring algorithm is employed to assign different scores to the proximity of domains based on the frequency of the closeness in the corpus. After developing the feature set, a naive Bayes multinomial (NBM) is used as a classifier. The experiment was evaluated on a database of 9000 text documents in Bangla acquired from various online news corpora, magazines and webpages from eight domains, namely Business, Travel, Food, State News, Entertainment, Sports, Medical and Science and Technology.

The rest of this chapter has the following structure. Related work is described briefly in Section 7.2, followed by the proposed methodology in Section 7.3, which involves discussion on data collection, pre-processing techniques, feature extraction and selection approach and classifiers being used. In Section 7.4, the results are analyzed along with acomparative study ofvarious classification algorithms and a comparison of the proposed methodology to existing methods. In Section 7.5, the paper is concluded showing some future directions in this field.

## 7.2  Existing Work

From the literature survey it can be observed that text categorization in English has gained great attention from researchers. A few works are briefly discussed here. DeySarkar et al. [1] worked based on a clustering technique using a naive Bayes classifier to classify documents from 13 datasets, obtaining quite an encouraging outcome. Guru and Suhil [2] evaluated their experiment on the 20Newsgroups dataset using support vector machine (SVM) and K-nearest neighbour (K-NN) algorithms based on the Term_Class relevance method. A bag-of-embeddings model was presented by Jin et al. [3] along with a stochastic gradient descent (SGD) classifier for classifying documents of the Reuters-21578 and 20Newsgroups datasets. Wang and Zhang [4] introduced two new schemes by including inverse category frequency (ICF) into TF-ICF and ICF-based methods as a feature extraction scheme for text categorization. Tellez et al. [5] presented a moderate and multi-propose classification algorithm that manages text categorization tasks effectively, irrespective of category and language. They defined their approach as a micro text classification model comprised of various text representations, transformations and a supervised learning algorithm used to easily implement them for categorization. Their approach proves to be effective for texts written informally. Linh et al. [6] presented a classification technique that uses the Dirichlet functions for directional data. Labeled training text documents have been used to automatically determine the number of topics for each text domain that is formed, suitable and marked. Their observation shows the advantages of their methodology in respect of recognizability, interpretability and capability for categorization of high dimensional and complex scattered databases.

Besides those in English, various works that have been carried out for Chinese, Japanese and Persian languages can be referred to here. To name a few, Wei et al. [7] used a Chinese corpus consisting of 14,150 Chinese text documents from 12 classes and conducted experiments based on three factors: (i) comparison of the performance among absolute and relative text frequency and N-gram frequency; (ii) comparison between sparseness and feature correlation; and (iii) performance comparison among 0/1 logical value, N-gram frequency numeric value and TF-IDF value. Wenliang et al. [8] carried out their experiment on a standard dataset of 14,459 texts from 37 domains using document frequency and chi square statistics to establish the class distribution of tokens. They have also used a global clustering model, referred to as a global CM, for Chinese text categorization. Mikawa et al. [9] introduced an efficient approach for Japanese text classification based on an extended cosine measure for classification of the Mainichi newspaper and customer reviews. The approach determines the distance between two documents based on a vector space model. In the case of Persian text categorization, Parvin et al. [10] categorized text documents from five text categories using normalized mutual information as well as a thesaurus and achieved a maximum accuracy of 81.45% for the value K = 1.

Besides English, Chinese, Japanese and Persian, several techniques of text categorization have been explored for Arabic. To mention a few, Al-Tahrawi [11] classified 1500 news articles from the Al Jazeera Arabic News corpus that were uniformly scattered into five domains using the chi square method. Al-Radaideh and Al-Khateeb [12] proposed a classification algorithm based on associative rules for classifying Arabic medical documents; Haralambous et al. [13] implemented light stemming and rootification procedures along with the TF-IDF and dependency grammar features on the dataset obtained from the Kalimat corpus; Ali and Ijaz [14] provided a comparative study of the statistical methods for classifying Urdu text documents using NB and SVM. Mainly, they pre-processed the documents using language-specific methods for developing a standard yet less dimensional feature vector. From the experiments, it can be seen that the performance increases when the value of n increases upto 3, but decreases when the values reaches 4.

In recent times, a few research works have been carried out on text classification for Indian languages as well. Some of these are briefly discussed here. For instance, Gupta and Gupta [15] combined NB- and ontology-based techniques to classify 184 Punjabi text documents extracted from various news corpora covering seven sub-domains of sports. Their experiment has shown that the combined approach performs better compared to other approaches. For the classification of Tamil text documents, ArunaDevi and Saveeth [16] used the C-feature extraction technique, which takes pairs of terms into consideration. For Marathi text document classification, Patil and Bogiri [17] presented an automatic Marathi text document categorization technique for categorizing 200 documents from 20 categories based on user profiles using the label induction grouping (LINGO) algorithm. In another experiment, Patil and Game [18] used four algorithms among which NB proved to be more effective in terms of accuracy and time for classifying Marathi text documents. Classification of texts using a language-specific WordNet for Assamese and Sanskrit were performed by Sarmah et al. [19] and Mohanty et al. [20]. In the case of Assamese [19], four classes were considered and an encouraging result of 90.27% accuracy was obtained for the same.

However, for Bangla, the evolution of text classification approaches has happened at a sorry pace. The lack of information in this field shows that there is a long way to go before we notice the urgent need for more research on this language. Some of the works performed on Bangla are reported here. The N-gram method was used by Mansur et al. [21] for categorizing one year of data from the Pratham Alo of Bangla news corpus. In another experiment, Mandal and Sen [22] experimented with four supervised learning methods to classify 1000 labelled texts into five categories and achieved a maximum accuracy of 89.14% with the SVM method. On the other hand, a SGD classifier was used by Kabir et al. [23] to classify text documents from nine distinct classes and achieved an accuracy of 93.85%, which is better compared to the previous one. In another work [24], an accuracy of 92.57% was observed based on a TF-IDF and SVM algorithm for classifying Bangla text documents of 12 text domains. Dhar et al. [25] used cosine similarity and Euclidean Distance based on a TF-IDF weighting scheme to categorize 1,000 web text documents from five domains: Business, State, Medical, Sports and Science. Their results show that the maximum accuracy (95.80%) was achieved with a Cosine Similarity measure. In another work, Dhar et al. [26] proposed a new technique of reduction, taking the top 40% of terms into account from a total set extracted from the training text documents based on a TF-IDF approach. The experiment was performed on 1960 text documents from five domains and achieved encouraging results using a LIBLINEAR classifier. The brief survey presented here certainly shows that we need to go a long way to develop an automatic text categorization system for Bangla language users, in order to serve their purposes efficiently.

## 7.3 Proposed Methodology

An overview of the proposed methodology is illustrated in Figure 7.1. On a feature set acquired using the feature extraction approach (UBoWs and the proximity between the text categories), a naive Bayes multinomial classification algorithm has been implemented to classify 9000 Bangla text documents obtained from various web sources. The details of the experimental results are presented in Section 7.4.

### 7.3.1 Data Collection

Both in terms of quality and quantity, data acts as an important factor in the results of any experiment. Developing a database requires much attention so that errors do not come up during the experiment and the appropriate outcome may be attained. However, from the brief literature survey, it has been

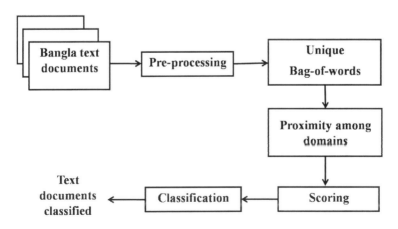

FIGURE 7.1
Block diagram of the proposed method.

**TABLE 7.1**

Text DocumentDistribution for Each Domain

| Domain | Number of Documents |
| --- | --- |
| Business | 1100 |
| Entertainment | 1100 |
| Food | 1000 |
| Medical | 1000 |
| State News | 1400 |
| Sports | 1400 |
| Science and Technology | 1000 |
| Travel | 1000 |
| **Total** | **9000** |

noticed that the domain-based Bangla text documents are not available in an adequate amount, so we had to build our own database to carry out the task. The text documents involved in the experiment were acquired from various online Bangla news corpora, magazines and webpages covering eight text domains: Business (B), Sports (SP), State News (P), Entertainment (E), Medical (M), Food (F), Science and Technology (ST) and Travel (T), as stated by leading newspapers. The distribution of text documents for each of the eight domains are provided in Table 7.1. The web sources of the text documents are: anandabazar.com, bartamanpatrika.com, ebela.in, deshirecipe.wordpress.com, ebanglarecipe.com, recipebangla.com, banglaadda.com, amaderchhuti.com, ebanglatravel.com.

### 7.3.2 Pre-Processing

Linear sequences of representations such as characters, words and phrases are used to form digital text documents. For a given sequence of characters in its respective text document, tokenization needs to be carried out before the application of the feature extraction and selection schemes. Tokenization is the task of segregating the sentences into pieces, called tokens. Without chopping the sentences clearly into tokens, further processing is not possible. Here, the splitting of sentences was performed based on a 'space' delimiter, with the total number of words being retrieved from the dataset being 2,806,159. The tokenized data speed up the processing of the system, make it more efficient and reduce the burden on system resources. This proves to be an advantage in systems relying on better performance. Since not all the words extracted from the dataset are relevant or informative enough to be in the feature set, the removal of stopwords from this set is necessary in order to filter the data. In the present experiment, punctuations, pronouns, English equivalent words, postpositions, English and Bangla numerals, interjections, some adjectives, conjunctions, some adverbs and all the articles have been considered stopwords, and the number of tokens acquired is 1,310,119.

Consideration of stopwords depends on a specific language as well as a specific problem.

### 7.3.3 Feature Extraction and Selection

There are a number of techniques used in text categorization. One of the most common and widely used approaches is called a 'bag of words'. This method is basically effortless to represent and accomplish, yet in NLP a simple method takes a long time to produce an encouraging outcome. The approach is also flexible enough to be used in innumerable procedures for extracting features from texts. The bag of words can be described as the occurrence of words within a text document. It is a procedure of expressing textual knowledge with machine learning algorithms. In natural language processing, the vectors are developed from text data that represent different linguistic characteristics of the texts. The approach only considers whether the particular term is present in the text document, not where in that document. In this approach, a histogram of the tokens within the text is considered a feature. The concept is that a text document is said to be similar to another text document if both have similar content. The bag-of-words model can be very simple or complex depending on the problem. The complex scenario arises while determining the design of the vocabulary of the tokens and the procedures of scoring the presence of those tokens.

In the present experiment, a UBoWs was developed and the proximity between two text categories was considered for the extraction of features, which could then be used for training the model classifier. Each distinct token in the UBoWs set represents a descriptive feature. The bag-of-words method has proved to be an achievement in tasks like language modeling and text document categorization. As is normally done in a bag-of-words model, all the words in every text document were considered, and then the frequencies of each token were counted. After getting the number of occurrences of each term, a certain number of tokens appearing more often were chosen. Prior to the implementation of machine learning algorithms, the raw texts needed to be converted into a vector of numbers. The weight of each unique term was determined by Equation 7.1. The percentage of unique terms of all text categories/domains are presented in Table 7.2. The trend of the features for the eight distinct text classes considered for the present experiment is graphically depicted in Figure 7.2. Scoring was performed to determine the level of proximity of a text document to a text category using Equation 7.2. Table 7.3 provides the percentage of proximity among different domains, estimated by considering the overlapping of unique terms among text documents of the domains considered in the present experiment.

$$W\left(t_x\right) = \left(1 + \log_2(1 + O_{t_x})\right) * O\left(t_x, d\right) \tag{7.1}$$

**TABLE 7.2**

Percentage of Unique Terms in All Text
Categories/Domains

| Domains | Unique Terms (%) |
|---|---|
| Business | 36.34 |
| Entertainment | 49.49 |
| Food | 73.42 |
| Medical | 51.28 |
| State News | 38.54 |
| Sports | 44.67 |
| Science & Technology | 51.72 |
| Travel | 65.44 |

where $W(t_x)$ is the weight of the unique term $t_x$ in a document, $O_{t_x}$ is the count of the presence of a term in the dataset and $O(t_x, d)$ is the count of the presence of a term in its respective document.

$$P(c) = p + W(t_x) \tag{7.2}$$

where $P(c)$ is the proximity level of a text document to a text category and p is the percentage of proximity among different domains. The normalized features have been selected from the feature set to get a more precise outcome from the automatic Bangla text categorization system.

### 7.3.4 Classifiers

The feature set being developed was used to train the NBM classification algorithm in order to classify Bangla text documents. The reason behind the selection of this classifier was its encouraging achievements in earlier text classification works [27–30]. The NBM algorithm is an extended version of naive Bayes that generally works better according to the kind of text documents. It generally estimates the probability of a domain in a document. The probability of a domain $\{P(v|w)\}$ is determined using Equation 7.3, where $\{P(t|v)\}$ is the conditional probability of term $t$ and $t_w$ represents the count of terms in a document $w$.

$$P(v|w) = P(v) \prod_{1 \le l \le t_w} P(t|v) \tag{7.3}$$

This equation determines the count of the tokens and manages the computations in actions. This extended version is basically used for the presence of a term multiple times in a categorization task. It computes the conditional probability $\{P(t|v)\}$ that denotes the relative frequency of a particular presence of a term $t$ for a given text category. It is evaluated using Equation 7.4, where $t$ represents a term of domain $v$, $N$ denotes the dataset, $B_w$ is the bag

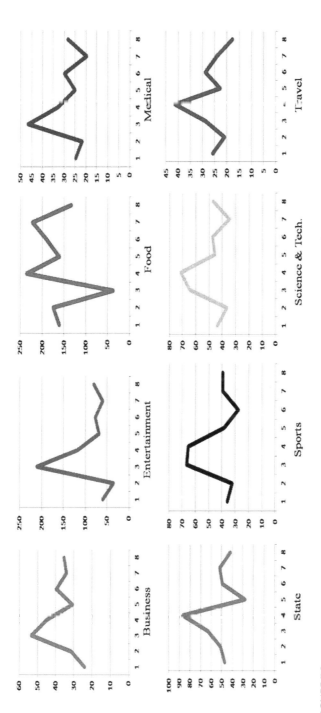

**FIGURE 7.2**
Feature trends of eight text categories.

**TABLE 7.3**

Percentage of Proximity among Different Domains

|     | B     | E     | F     | M     | P     | SP    | ST    | T     |
|-----|-------|-------|-------|-------|-------|-------|-------|-------|
| B   | —     | 50.64 | 35.60 | 64.84 | **81.64** | 46.97 | 74.66 | 45.08 |
| E   | 50.64 | —     | 16.97 | 32.18 | **65.70** | 55.00 | 29.00 | 65.18 |
| F   | 35.60 | 16.97 | —     | 24.07 | 17.30 | 16.90 | 17.89 | **64.28** |
| M   | 64.84 | 32.18 | 24.07 | —     | **76.04** | 40.13 | 73.94 | 24.16 |
| P   | **81.64** | 65.70 | 17.30 | 76.04 | —     | 63.18 | 28.82 | 23.95 |
| SP  | 46.97 | **55.00** | 16.90 | 40.13 | 63.18 | —     | 26.35 | 41.13 |
| ST  | **74.66** | 29.00 | 17.89 | 73.94 | 28.82 | 26.35 | —     | 20.63 |
| T   | 45.08 | **65.18** | 64.28 | 24.16 | 23.95 | 41.13 | 20.63 | —     |

of words and $\{N_{vt}\}$ determines the presence of term $t$ in documents from domain $v$, taking multiple presences of a word into consideration.

$$P(t|v) = \frac{N_{vt}}{\sum_{t' \in B_w} N_{vt'}} \qquad (7.4)$$

## 7.4 Results and Analysis

The results obtained by carrying out Bangla text categorization on 9,000 documents from eight categories are presented in this section. For this purpose, WEKA [31], a highly accepted open source classification tool, was used for the implementation of the classifier considered here. K fold cross-validation has been performed on the obtained dataset, where K was chosen to be 10 [32]. The maximum accuracy being achieved is 97.30% over 1,000 iterations using the NBM classification algorithm; the confusion matrix obtained is presented in Table 7.4. Experiments were also performed over different

**TABLE 7.4**

Confusion Matrix Obtained Using NBM

|     | B    | E    | F   | M   | P    | SP   | ST  | T   |
|-----|------|------|-----|-----|------|------|-----|-----|
| B   | 1024 | 8    | 0   | 22  | 42   | 4    | 0   | 0   |
| E   | 0    | 1079 | 0   | 0   | 17   | 4    | 0   | 0   |
| F   | 0    | 0    | 990 | 4   | 0    | 0    | 0   | 6   |
| M   | 7    | 0    | 0   | 968 | 16   | 0    | 9   | 0   |
| P   | 19   | 4    | 0   | 12  | 1363 | 2    | 0   | 0   |
| SP  | 4    | 12   | 0   | 0   | 8    | 1372 | 0   | 4   |
| ST  | 9    | 0    | 0   | 9   | 3    | 0    | 979 | 0   |
| T   | 0    | 10   | 6   | 0   | 0    | 2    | 0   | 982 |

cross-validation folds as well as for varying numbers of iterations using the best-performing cross-validation folds. The obtained results are presented in Tables 7.5 and 7.6. The domain-wise accuracy based on various parameters of the system's performance is provided in Figure 7.3.

Based on the confusion matrix obtained using NBM with 10-fold cross-validation over 1000 training iterations, the average percentages of correctly classified text documents as well as the rate of misclassification for

**TABLE 7.5**

Performance of NBM for Different Cross-Validation Folds

| Folds | 5 | 10 | 15 | 20 |
|---|---|---|---|---|
| Accuracy (%) | 96.12 | **97.30** | 95.39 | 95.24 |

**TABLE 7.6**

Performance of NBM for Different Training Iterations Using TenCross-Validation Folds

| Iterations | 500 | 1000 | 1500 | 2000 |
|---|---|---|---|---|
| Accuracy (%) | 95.98 | **97.30** | 96.42 | 95.47 |

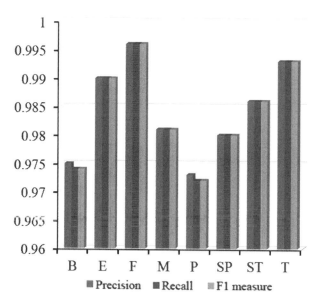

**FIGURE 7.3**
Domain-wise evaluation in terms of precision, recall and $F_1$ measure.

**TABLE 7.7**

Average Percentage of Correct and Incorrect Classifications

| Domains | Correct Classification (%) | Incorrect Classification (%) |
|---|---|---|
| Business | 93.10 | 6.90 |
| Entertainment | 98.10 | 1.90 |
| Food | 99.00 | 1.00 |
| Medical | 96.80 | 3.20 |
| State News | 97.36 | 2.64 |
| Sports | 98.00 | 2.00 |
| Science and Technology | 97.90 | 2.10 |
| Travel | 98.20 | 1.80 |
| **Average** | **97.31** | **2.69** |

each domain are given in Table 7.7, which is calculated using the following equations:

$$\text{Classification} = \frac{\text{Correctly classified documents}}{\text{Total number of text documents}} * 100\%. \qquad (7.5)$$

$$\text{Misclassification} = \frac{\text{Incorrectly classified documents}}{\text{Total number of text documents}} * 100. \qquad (7.6)$$

The result achieved from the experiment has also been compared with other commonly used classification algorithms, namely multilayer perceptron (MLP) [33], SVM, random forest (RF) [34], rule based (PART) and LogitBoost (LB) to test the robustness of our system. The comparative analysis is shown in Table 7.8, from which it can be seen that the methodology adopted in the present experiment performs better using NBM in comparison with others in the case of Bangla text categorization. The worst was performed by SVM; MLP, RF, LB and PART obtained the 2nd, 3rd, 4th and 5th ranks, respectively.

**TABLE 7.8**

Comparison of Classifiers

| Classifiers | Accuracy (%) |
|---|---|
| NBM | **97.30** |
| MLP | 96.89 |
| SVM | 94.21 |
| RF | 96.19 |
| PART | 94.43 |
| LB | 95.98 |

## 7.5 Conclusion

This chapter presents an approach that uses the NBM classification algorithm based on a feature extraction scheme (unique bag of words and proximity among domains) to classify Bangla text documents into their respective text categories. The results were also compared with other commonly used classifiers and it has been observed from the experiments that the NBM performs better compared to other algorithms for Bangla text classification. In future, this experiment can be expanded to include a larger number of domain-wide Bangla text documents, as well as more domains or text categories. Also, new hybrid approaches can be introduced to classify text documents into their predetermined text categories. We also plan to experiment with other machine learning techniques like those based on deep learning and active learning [35] approaches to obtain better results. In future, this study can be used on the datasets of other languages as well.

## Acknowledgment

One of the authors thanks DST for its support in the form of an INSPIRE fellowship.

## References

1. DeySarkar, S., Goswami, S., Agarwal, A., Aktar, J.: A Novel Feature Selection Techniquefor Text Classification Using Nave Bayes, *International Scholarly Research Notices*, 2014, p. 10, 2014.
2. Guru, D., Suhil, M.: A Novel Term Class Relevance Measure for Text Categorization, *Proc. of ICACTA*, pp. 13–22, 2015.
3. Jin, P., Zhang, Y., Chen, X., Xia, Y.: Bag-Of Embeddings for Text Classification, Proc. of IJCAI, pp. 2824–2830, 2016.
4. Wang, D., Zhang, H.: Inverse-Category-Frequency Based Supervised Term Weighting Schemes for Text Categorization, *Journal of Information Science and Engineering*, 29, pp. 209–225, 2013.
5. Tellez, E., Moctezuma, D., Miranda-Jimnez, S., Graff, M.: An Automated Text Categorization Framework Based on Hyperparameter Optimization, *Knowledge-Based Systems*, 149, pp. 110–123, 2018.
6. Linh, N. V., Anh, N. K., Than, K., Dang, C. N.: An Effective and Interpretable Method for Document Classification, *Knowledge and Information Systems*, 50(3), pp. 763–793, 2017.

7.  Wei, Z., Miao, D., Chauchat, J., Zhao, R., Li, W.: N-Grams Based Feature Selection and Text Representation for Chinese Text Classification, *International Journal of Computational Intelligence Systems*, 02, pp. 365–374, 2009.

8.  Wenliang, C., Xingzhi, C., Huizhen, W., Jingbo, Z., Tianshun, Y.: Automatic Word Clustering for Text Categorization Using Global Information, Proc. of AIRS, pp. 1–11, 2015.

9.  Mikawa, K., Ishidat, T., Goto, M.: A Proposal of Extended Cosine Measure for Distance Metric Learning in Text Classification, Proc. of ICSMC, pp. 1741–1746, 2011.

10. Parvin, H., Dahbashi, A., Parvin, S., Minaei, B.: Improving Persian Text Classification and Clustering Using Persian Thesaurus, Proc. of ICDCAI, pp. 493–500, 2012.

11. Al-Tahrawi, M.: Arabic Text Categorization Using Logistic Regression, *International Journal of Intelligent Systems and Applications*, 06(6), pp. 71–78, 2015.

12. Al-Radaideh, Q.A., Al-Khateeb, S.S.: An Associative Rule-Based Classifier for Arabic Medical Text, *International Journal of Knowledge Engineering and Data Mining*, 03, pp. 255–273, 2015.

13. Haralambous, Y., Elidrissi, Y., Lenca, P.: Arabic Language Text Classification Using Dependency Syntax-Based Feature Selection, Proc. of ICALP, p. 10, 2014.

14. Ali, A., Ijaz, M.: Urdu Text Classification, Proc. of ICFIT, pp. 1–7, 2009.

15. Gupta, N., Gupta, V.: Punjabi Text Classification Using Naive Bayes, Centroid and Hybrid Approach, Proc. of SEANLP, pp. 109–122, 2012.

16. ArunaDevi, K., Saveeth, R.: A Novel Approach on Tamil Text Classification Using C-Feature, *International Journal of Scientific Research & Development*, 02, pp. 343–345, 2014.

17. Patil, J.J., Bogiri, N.: Automatic Text Categorization Marathi Documents, *International Journal of Advanced Research in Computer Science and Management Studies*, 03, pp. 280–287, 2015.

18. Patil, M., Game, P.: Comparison of Marathi Text Classifiers, ACEEE, *International Journal on Information Technology*, 04, pp. 11–22, 2014.

19. Sarmah, J., Saharia, N., Shikhar, K.: A Novel Approach for Document Classification Using Assamese Wordnet, Proc. of IGWC, pp. 324–329, 2012.

20. Mohanty, S., Santi, P., Mishra, R., Mohapatra, R., Swain, S.: Semantic Based Text Classification Using Wordnets: Indian Language Perspective, Proc. of ICECCE, pp. 321–324, 2006.

21. Mansur, M., UzZaman, N., Khan, M.: Analysis of N-Gram Based Text Categorization for Bangla in a Newspaper Corpus, Proc. of ICCIT, p. 08, 2006.

22. Mandal, A.K., Sen, R.: Supervised Learning Methods for Bangla Web Document Categorization, *International Journal of Artificial Intelligence & Applications*, 05(5), pp. 93–105, 2014.

23. Kabir, F., Siddique, S., Kotwal, M.R.A., Huda, M.N.: Bangla Text Document Categorization Using Stochastic Gradient Descent (SGD) Classifier, Proc. of ICCCIP, pp. 1–4, 2015.

24. Islam, Md.S., Jubayer, F.E.Md., Ahmed, S.I.: A Support Vector Machine Mixed with TF-IDF Algorithm to Categorize Bengali Document, Proc. of ICECCE, pp. 191–196, 2017.

25. Dhar, A., Dash, N.S., Roy, K.: Classification of Text Documents through Distance Measurement: An Experiment with Multi-Domain Bangla Text Documents, Proc. of ICACCA, p. 16, 2017.

26. Dhar, A., Dash, N.S., Roy, K.: Application of TF-IDF Feature for Categorizing Documents of Online Bangla Web Text Corpus, Proc. of FICTA, pp. 51–59, 2017.

27. Kibriya, A.M., Frank, E., Pfahringer, B., Holmes, G.: Multinomial Naive Bayes for Text Categorization, Revisited, Proc. of AJCAI, pp. 488–499, 2005.

28. Kim, S.B., Han, K.S., Rim, H.C., Myaeng, S.H.: Some Effective Techniques for Naive Bayes Text Classification, *IEEE Transactions on Knowledge and Data Engineering*, 18(11), pp. 1457–1466, 2006.

29. Rehman, A., Javed, K., Babri, H.A.: Term-Weighting Approaches in Automatic Text Retrieval, *Information Processing and Management*, 53(2), pp. 473–489, 2017.

30. Wilbur, J.W., Kim, W.: The Ineffectiveness of Within-Document Term Frequency in Text Classification, *Information Retrieval*, 12(5), pp. 509–525, 2009.

31. Hall, M., Frank, E., Holmes, G., Pfahringer, B., Reutemann, P., Witten, I.H.: The Weka Data Mining Software: An Update, *ACM SIGKDDexplorations Newsletter*, 11(1), pp. 10–18, 2009.

32. Obaidullah, Sk.M., Santosh, K.C., Halder, C., Das, N., Roy, K.: Automatic Indic Script Identification from Handwritten Documents: Page, Block, Line and Word-Level Approach, *International Journal of Machine Learning and Cybernetics*, pp. 1–20, 2017.

33. Obaidullah, Sk.M., Halder, C., Santosh, K.C., Das, N., Roy, K.: PHDIndic_11:Page-Level Handwritten Document Image Dataset of 11 Official Indic Scripts for Script Identification, *Multimedia Tools and Applications*, 77, pp. 1643–1678, 2017.

34. Obaidullah, Sk.M., Goswami, C., Santosh, K.C., Das, N., Halder, C., Roy, K.: Separating Indic Scripts with `Matra' for Effective Handwritten Script Identification in Multi-Script Documents, *International Journal of Artificial Intelligence & Pattern Recognition*, 31, 1753003, 2017.

35. Bouguelia, M.R., Nowaczyk, S., Santosh, K.C., Verikas, A.: Agreeing to Disagree: Active Learning with Noisy Labels without Crowdsourcing, *International Journal of Machine Learning and Cybernetics*, 9, p. 13, 2017.

# 8

## Supervised Learning for Aggression Identification and Author Profiling over Twitter Dataset

**Kashyap Raiyani and Roy Bayot**

**CONTENTS**

## 8.1 Introduction

An individual document has its own unique way of representation. Based on this illustration, there are approaches to identifying the author's gender and the aggression behind the writing. This knowledge is crucial when dealing with forensic science, cyberbullying and hate speech detection, for example, where knowing the profile of a writer could be of fundamental interest. In cases similar to forensic linguistics, being able to comprehend the linguistic profile of a suspicious text message (a style used by a particular writer) and identify characteristics (expression as evidence) just by analyzing the text would certainly assist in considering suspects. Similarly, from a marketing viewpoint, companies may be interested in knowing what class of people like or dislike their products and how aggressively they talk about them. Blogs and online reviews can aid in retrieving this information. However, in most instances, without a proper system, this knowledge cannot be retrieved automatically. Fortunately, with the help of machine learning or supervised learning, it is possible to train a system that has been provided with the right dataset to automatically classify/identify the gender and aggression. Even though it is machine-generated data, regardless of the trust issue, this erudition could assist in manually cross-verifying the obtained values. For this practice, a separate model was trained using a social media (Facebook and Twitter) dataset for gender and aggression identification. To measure the performance of the models, they were submitted in two shared tasks, namely Task 1: Aggression Identification and Task 2: Author Profiling. The results of the models exceed the average performance in automated identi-fication of a writer's gender and of the aggression level. These models can work with various languages like Arabic, English, Hindi and Spanish. The principal concepts used were data/text analysis, preprocessing, metavari-able selection and system development. Extra research leads to acquiring words that are unseen to the model, making the model adaptive.

## 8.2 Overview of Aggression Identification

Text mining and the acquisition of meaningful and relevant information from text corpora have become an important field for research and compa-nies where the decision-making support system is the priority. Table 8.1 dis-cusses recently published research on aggression and hate speech language identification.

### 8.2.1 Dataset

For any machine learning approach, the dataset plays an important role. The dataset used here was obtained from the shared task on "Aggression Identification" [11] of a workshop named "Trolling, Aggression and Cyberbullying" [12]. The dataset had 15,000 Facebook posts and comments written in Hindi (in both Roman and Devanagari script) and English. They were annotated with three labels, namely "Overtly Aggressive (OAG)", "Covertly Aggressive (CAG)" or "Non-aggressive (NAG)". The idea here was to predict the label of an unknown post or comment using supervised learning. Before any further processing, data was visualized for better

**TABLE 8.1**

Related Work

| Reference | Feature | Method | Observation |
|---|---|---|---|
| [1] | Bag of words, unigram, bigram, trigram and character-level n-gram | Word generalization | "Empathized on knowledge-based features and meta-information." |
| [2] | Character n-grams, word n-grams and word skip-grams | Character 4-grams | "Challenge lies in discriminating profanity and hate speech from each other." |
| [3] | Binary classification | Using binary classified (unigram) created offensive words vocabulary | "Able to achieve 86% accuracy in detecting offensive vocabulary." |
| [4] | Lexicon, linguistic, n-grams, syntactic, word2vec | Developed abusive language label corpus | "Model outperforms classical/state-of-art blacklisting, regular expression and NLP models." |
| [5] | Lexicons, bag of words, clustering | Distributional semantics used to find word context or cooccurrence | "Used features are not enough to precisely classify hate speech and to detect hate speech from offensive speech." |
| [6] | Character 4-grams, word vectors and word vectors combined with character n-grams | Convolutional neural networks and max pooling | "Word2vec embeddings outperformed all others." |
| [7] | Word n-grams and word skip-grams | Convolutional neural networks – gated recurrent networks with word-based features | "Captured both word sequence and order information in short texts and were able to set new benchmark." |
| [8] | Labels used included Offensive, Abusive, Hate Speech, Aggressive Behavior, Cyberbullying, Spam and Normal | 80,000 tweets were annotated and made publicly available | "They were able to classify whether the state of behavior is Hateful or Normal." |
| [9] | Binary classifier | Breaking data into sub-categories until it becomes binary | "Binary classification gives better results compared to the multi-label classifier." |
| [10] | Part of speech (POS), grammatical dependencies and hate speech keywords | Bayesian logistic regression, support vector machine and random forest decision tree were used | "Ensemble classification or stack classification approach are most suitable for classifying cyber hate, given the current feature sets." |

understanding and system modeling. There are two data files for each language.

1. "en_train/hi_train.csv" contains the train set, with 12,000 records. (Overtly Aggressive = 2708, Covertly Aggressive = 4240 and Non-aggressive = 5052).

2. "en_dev/hi_dev.csv" contains the test set with 3001 records. (Overtly Aggressive = 711, Covertly Aggressive = 1057 and Non-aggressive = 1233).

The approach/design of the model depends upon data characteristics and the features data holds. The data characteristics are discussed in Section 8.2.2, followed by preprocessing and feature extraction.

## 8.2.2 Data Characteristics

Tweets are mostly extracted using their API, and the extractor defines the ranges/properties like geographical region, language and topic of interest. In context with gender identification and aggression identification, the most common data characteristics to consider are the following:

- The region of data collection: With the help of this information it becomes easy to predict/remove unwanted noise in the dataset. For example, understanding of slang language, common traits of user behavior/sarcasm.

- Usage of language-specific stop-words: In general, it is observed that 30% of user posts contain common words. For instance, if tweets are in English, then these words would be the, is, a, an and so on.

- The modality of words in terms of abbreviation/jargon: Mostly, in informal communication, people tend to use abbreviations or jargon instead of writing complete words. For example, ROFL, LOL and many more.

- Usage of emojis and their impact on tweets: Apparently, emojis give more insight than the text or can add extra meaning to the tweet. Mainly, the focus here is to check the distribution level and how much noise an emoji adds to the dataset.

- In cases of supervised leering, class/label distribution over the dataset: It is preferable to have an equally distributed dataset for machine learning approaches, but many datasets suffer from class imbalance.

- Usage of hashtags: As it is a Twitter and Facebook dataset, it contains several hashtags and joint words. Before doing any preprocessing to it, one needs to consider how much meaning these hold in context.

Table 8.2 shows the data characteristics of the dataset.

**TABLE 8.2**

Dataset Characteristics

| Region of Data Collection | India |
|---|---|
| English stop-words | High |
| Modality abbreviation/jargon | High |
| Usage of emojis | Distributed over classes |
| Class/label distribution | Imbalance |
| Usage of hashtags | High |

## 8.2.3 Data Preprocessing

Data preprocessing is the most important phase of any system modeling and, generally, data characteristics are considered for preprocessing. Here, for the previously mentioned data characteristics, tweet preprocessing is discussed.

- Stem words: As per language morphology. A stem word is a word without affixes. Removing a word from, or making it into, its stem form does not affect its semantic meaning. In addition, it reduces the number of distinct words in the dataset. This might be very helpful when a word dictionary is created using a dataset. The rule is as follows:
- Rule: regexp = '^(.*?)(ing)(ed)(ment)(ion)?$'
- Stop-words: The removal of stop-words helps in many cases where the semantic meaning of the sentence is not captured. Again, it helps in reducing the sentence length and removing distributed language noise by only keeping meaningful words in the sentence. On the other side, it is not advisable to remove stop-words when the user wants to capture the (semantic) meaning of the sentence.
- Symbolic emojis: In most textual social media communication, emojis are used to represent the emotion of the author. Normally, all text platforms have emojis (an image), but most people use the symbolic emoji. To do this, the following regular expressions are used:
- "(:\s?D|:-D|x-?D|X-?D)" for # Laugh -- :D, : D, :-D, xD, x-D, XD, X-D
- "(:\s?\)|:-\)|\(\s?:|\(-:|:\'\))" for # Smile -- :), : ), :-), (:, ( :, (-:
- "(<3|:\*)" for # Love -- <3, :*
- "(;-?\)|;-?D|\(-?;)" for # Wink -- ;-), ;), ;-D, ;D, (;, (-;
- "(:\s?\(|:-\(|\)\s?:|\)-:)" for # Sad -- :-(, : (, :(, ):, )-:
- "(:,\(|:'\(|:"\()" for # Cry -- :,(, :'(, :"(
- Abbreviation/jargon: The abbreviation/jargon depends upon the region of data collection and, accordingly, regular expressions were used to replace such words with their original form.

- Segmentation: The most commonly used character "#" followed by many words. In most of the cases, research does not use this information but here, ekphrasis [13] is used to remove the segmentation and to find those following words.

Table 8.3 shows the preprocessing done over the dataset.

### 8.2.4 Feature Extraction

The key aspect of information retrieval is feature extraction. Depending upon the number of features extracted, one can define the system to use them. In general, the most common features extracted are character-level unigrams, bigrams and trigrams. Apart from these, TFIDF Vectorizer, Doc2Vec, Word2Vec, distributed bag of words (DBOW), distributed memory concatenated (DMC) and distributed memory mean (DMM) are also used. Here, TFIDF is used to measure the term-weighting in the document and Doc2Vec and Word2Vec generate vectors for words/sentences/paragraphs/documents. DBOW, DMC and DMM are used as parts of Doc2Vec. In many systems, researchers also use part-of-speech (POS) tags as one of the features. Table 8.4 shows the feature extracted over the dataset.

**TABLE 8.3**

Preprocessing over Dataset

| Preprocessing/Language | English | Hindi |
| --- | --- | --- |
| Stem Words | ✓ | ✗ |
| Stop-Words | ✓ | ✗ |
| Symbolic Emojis | ✓ | ✓ |
| Abbreviations/Jargon | ✓ | ✓ |
| Segmentation | ✓ | ✓ |

**TABLE 8.4**

Extracted Features over Dataset

| Feature/Language | English | Hindi |
| --- | --- | --- |
| Unigram | ✓ | ✗ |
| Bigram | ✓ | ✗ |
| Trigram | ✓ | ✗ |
| TFIDF Vectorizer | ✓ | ✗ |
| Doc2Vec – DBOW | ✓ | ✗ |
| Doc2Vec – DMC | ✓ | ✗ |
| Doc2Vec – DMM | ✓ | ✗ |
| Word Dictionary | ✓ | ✓ |

### 8.2.5 Experimental Setup

To do the system modeling, two methods were used: (a) traditional machine learning or linear modeling and (b) deep learning. For the linear model, logistic regression was used; for deep learning modeling, Keras [14] was used as a front-end and Tensorflow [15] as a back-end. In the case of deep learning, a Glove [16] pre-trained word vector (glove.twitter.27B) was used. From the dataset (mentioned in Section 8.2.1), 9600, 2399 and 3001 samples are used for training, validating and testing, respectively.

### 8.2.6 System Modeling

It is always better to use both linear modeling and deep machine learning. Table 8.5 shows the results in terms of accuracy over logistic regression with a different feature set.

The most traditional deep learning algorithms are long short-term memory (LSTM), convolutional neural network (CNN) and recurrent neural network (RNN). Apart from these, Facebook FastText Text Classification is also used here. Table 8.6 presents the accuracy obtained in the English development dataset with the algorithms just mentioned. Here, the dataset is divided into three parts: (a) training, 9600 samples; (b) validating, 2399 samples; and (c) testing 3001 samples.

As a final model, a fully connected dense system architecture was designed. Here, instead of using pre-trained word vectors, a word dictionary was used. In this method, all of the words were indexed by an integer followed by its binary conversation as a word representation. Figure 8.1 shows the architecture.

Table 8.7 shows the parameter of the architecture.

Here, the activation function is the mathematical function which determines the flow from one end to the other. It acts as a filter: For example, Relu will turn all negative values to 0. On the other hand, sigmoid will allow values between 0 and 1, and softmax will distribute values between 0 and 1 over the number of nodes (in this case, three). The result of this model is discussed in the next subsection.

**TABLE 8.5**

Accuracy over Logistic Regression with Different Feature Sets

| Feature/Word Gram | Unigram | Bigram | Trigram |
| --- | --- | --- | --- |
| TFIDF Vectorizer | 56.91 | 56.49 | 56.99 |
| DBOW | 54.86 | 54.32 | 53.56 |
| DMC | 46.63 | 48.30 | 48.30 |
| DMM | 52.31 | 53.65 | 53.40 |
| DBOW + DMC | 54.48 | 53.73 | 53.69 |
| DBOW + DMM | 54.40 | 54.82 | 54.53 |

**TABLE 8.6**

Accuracy over Deep Learning Algorithms [17]

| Method | Acc on Test (%) |
| --- | --- |
| Single layer LSTM | 37.93 |
| Multilayer LSTM | 39.20 |
| Conv1D & GlobalMaxPooling1D | 37.37 |
| Conv1D & MaxPooling1D with 100 Hidden Layer | 37.73 |
| Convolutional Layers with LSTM | 39.03 |
| Convolutional Layers with Bidirectional LSTM | 37.53 |
| FastText Text Classification | 54.00 |
| FastText Text Classification with Skip Gram Model | 37.00 |
| FastText with Conv1D, MaxPooling1D & Bidirectional LSTM | 38.53 |
| Multiple Input RNN with Keras | 37.00 |
| Concatenate: 2 Bidirectional LSTM | 38.00 |
| Concatenate: Bidirectional with Conv1D & MaxPooling1D | 37.50 |

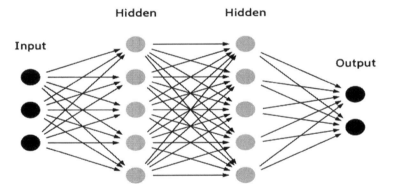

**FIGURE 8.1**
Dense architecture.

**TABLE 8.7**

Dense Architecture Parameters

| Layer | Nodes | Activation Function |
| --- | --- | --- |
| Input | Size of dictionary | n/a |
| 1st Hidden | 512 | Relu |
| 2nd Hidden | 256 | Sigmoid |
| Output | 3 | Softmax |

**TABLE 8.8**

Results: Dense vs. FastText [17]

| Model/Language | FB: English | FB: Hindi | Twitter: English | Twitter: Hindi |
|---|---|---|---|---|
| Dense | 58.13 | 59.51 | 60.09 | 48.30 |
| FastText | 57.53 | 58.38 | 55.44 | 45.28 |

### 8.2.7 Results

Four test set categories were used: (a) Facebook and (b) Twitter, English; and (c) Facebook and (d) Twitter, Hindi. Here, Facebook test data was used to check the level of supervised learning while Twitter test data was used to check the transfer learning. The dense model is compared to the FastText model. Table 8.8 shows the results of the submitted systems.

## 8.3 Overview of Author Profiling

One aspect of document processing is authorship identification, wherein a specific author is predicted given a certain text. Author profiling is a subsection where instead of giving the specific author, characteristics such as age and gender of the author are predicted. The profiling methods stem mainly from the different features extracted.

One paper that outlines various features is given by Stamatatos [18]. He outlines lexical, character, syntactic, semantic and application-specific features. These feature types were evident in submissions to the author profiling task of PAN, an ongoing project of the Conference and Labs of the Evaluation Forum (CLEF). There are various editions, but the ones pertinent to our task are PAN 2015 [19], whose task is to classify tweets for four languages, and PAN 2016 [20], whose task is to do a cross-genre evaluation. The main types of features extracted include style-based, content-based, n-gram–based, information retrieval–based and collocation-based. In PAN 2015, for instance, character n-gram models were used in [21, 22]. Word n-grams were also used in works such as those of [23–25]. TFIDF n-grams were used in [26, 27]. POS n-grams were also used in [26, 28]. Another set of features stem from word vectors. There are various word vector implementations proposed. One that is widely used is Word2Vec by Mikolov [29, 30]. The Word2Vec algorithm essentially has two modes: continuous bag of words and skip-grams. To obtain the word vectors, the program must initialize the vectors from random numbers. Then it reads from a large corpus such as Wikipedia. Using the sequence of words in the text given in the corpus, the program uses the initial random vectors that represent words to predict other words. In continuous bag of words, for instance, a target word is predicted using the

**TABLE 8.9**

Age Distribution for the PAN 2015 Dataset

| Language/Age | 18–24 | 25–34 | 35–49 | 50+ |
|---|---|---|---|---|
| English | 58 | 60 | 22 | 12 |
| Spanish | 22 | 56 | 22 | 10 |

surrounding context words. Skip-grams, on the other hand, use a word to predict its contextual words.

### 8.3.1 Datasets

There are two datasets coming from PAN. The first one is from the PAN 2015 dataset [19]. It is comprised of users with a varying number of tweets for four different languages: English, Spanish, Italian and Dutch. However, we only use the section for English and Spanish. The number of users differs. English has 152 users while Spanish has 110. The dataset was balanced based on gender and labels are given for gender, age and personality. There were four different categories of age: 18–24, 25–34, 35–49 and 50 and above. This is shown in Table 8.9. In terms of personality traits, each user has a score between –0.5 and 0.5 on the big five personality traits in the Five Factor Theory [31]. These include extroversion, emotional stability/neuroticism, agreeableness, conscientiousness and openness to experience.

The second dataset come from the PAN 2016 dataset [20]. It is comprised of tweets from different users in English and Spanish. There are 428 users for English and 250 for Spanish, each with a different number of tweets. The labels available are only age and gender. The dataset is balanced over gender and the age labels are different from the previous dataset. It is now categorized as 18–24, 25–34, 35–49, 50–64 and 64 and above. The age distribution is given in Table 8.10.

### 8.3.2 Preprocessing

There was not much preprocessing done. For each language, XML files from each user are read. Then the tweets taken from each user are extracted. Depending on the experimental setup, these tweets are either concatenated into one line to form one training example or retained as is to form one

**TABLE 8.10**

Age Distribution for the PAN 2016 Dataset

| Language/Age | 18–24 | 25–34 | 35–49 | 50–64 | 65+ |
|---|---|---|---|---|---|
| English | 28 | 140 | 182 | 80 | 6 |
| Spanish | 16 | 64 | 126 | 38 | 6 |

training example per tweet. The examples are then put into their lowercase equivalents. Twitter-specific features such as hashtags, numbers, mentions, shares and retweets are not removed, processed or transformed. Stop-words are also retained. The resulting file is used for feature extraction.

### 8.3.3 Feature Extraction

There were a few features extracted. These included a bag-of-words feature, word n-grams and part-of-speech n-grams. The experiments included two bag-of-word features. The first is a normalized term frequency, which was normalized by the number of terms in the training example. The second is TFIDF, which was normalized by the inverse document frequency. Word bigrams and trigrams were used in the experiments as well. There were also part-of-speech n-grams. This included only unigrams and bigrams. The tagger used was Schmid's TreeTagger program [32]. Another feature was the average of word vectors. We used a Gensim [33] implementation of Word2Vec to get word vectors with dimensions 100 and 300. We used the February 5, 2016 Wikipedia dump.

### 8.3.4 Experimental Setup

There were multiple things that needed to be tested. First, we wanted to know which of the conventional features would be indicative of the classes for age, gender and personality profiling. Then we wanted to observe if using word vectors could have a comparable or bigger improvement over classification.

Finally, we wanted to see if using deep learning architectures could also yield better results. Since there were two different datasets, also with different modalities of the classification algorithms, as well as different things to be tested, there were two setups created. This is illustrated in Figure 8.2. The first setup involved ten-fold cross-validation when first dealing with the PAN 2015 and PAN 2016 datasets. In this setup, one training example is the concatenation of tweets for each user. Established features such as bag of words, part-of-speech n-grams and word n-grams were used on PAN 2015 in conjunction with support vector machines [34] and various kernels. This allowed us to compare the effect of the features over this dataset. We then had the same cross-validation setup using the average of word vectors for PAN 2016 also in conjunction with support vector machines. This enabled us to observe the effect of such vectors on classification.

The second setup uses hold-out and the PAN 2016 dataset. In this setup, one training example is a tweet. The dataset was first split by users into 70% training and 30% test. Training examples were made from each tweet coming from the 70% portion. Test examples were also made from each tweet coming from the 30% portion. The training examples were then used in a convolutional neural network model based on Kim [35]. This enabled us to observe some factors for classification.

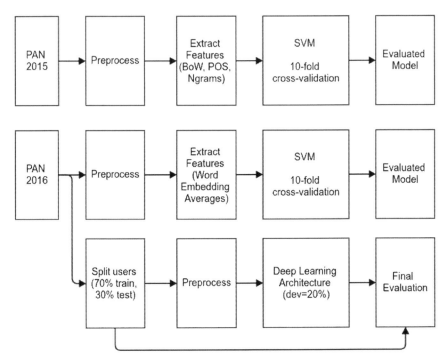

**FIGURE 8.2**
Illustration of the different setups.

### 8.3.5 Algorithm and Fine-Tuning the Model

There were two learning algorithms used for the experiments: support vector machines (SVM) and convolutional neural network (CNN). For the first part, we used an SVM with a linear kernel with the default penalty parameter of the error term to 1 to see which of the conventional features worked the best for this problem. A Wilcoxon signed rank test [36] was then done to check if the difference between the results was significant or not.

Then, for word vector averages, we experimented with a polynomial kernel and a radial basis function kernel. We experimented with polynomial degrees of 1, 2 and 3 with the penalty parameter of the error term to be either 0.01, 1 or 100. We also experimented with radial basis function kernels with gammas of either 0.01, 1 or 100 and with the same options of the penalty to the error term as the previous. We did not use statistical tests in this section.

The convolutional neural network that we used follows that of Kim [35] which is comprised of an embedding layer, followed by a convolutional layer with three filters of window sizes 3, 4 and 5, wherein each filter window has 100 feature maps. We used the tanh function for the non-linearity. After the feature maps, max-over-time pooling was performed. Then a dropout layer was added. Our dropout probability was 0.5. Then everything was collected for the final softmax layer with weight vectors constrained to $l_2$ norm. Everything was trained using a stochastic gradient descent over shuffled mini batches using Adadelta. Each mini-batch had 3000 examples. With this setup, we experimented with varying between 100- and 300-word vector dimensions using skip-grams as well as keeping the word vectors static or not static. Keeping the vectors static means that we would not change the vector values during training. Furthermore, since the dataset is in users while the model is in tweet classification, we aggregated the classification result for the tweets belonging to the same user and used a majority vote to indicate the class for each user. We would also be able to see the effect of using the majority vote on the classification.

### 8.3.6 Results

Our first set of experiments involved checking which feature types give the best classification accuracy for PAN 2015. The results for all the languages and tasks are explained in greater detail in the paper in [37]. The succeeding tables, however, are for English and Spanish age and gender classification as well as a regression on personality traits.

Gender classification accuracy is shown in Table 8.11. Between bag-of-words features, TFIDF features gave better results than normalized term frequency. Between word n-gram features, word trigrams worked better than word bigrams for English, but the converse worked better for Spanish. Among POS n-grams, the combination of POS unigrams and bigrams worked well for both English and Spanish. However, looking at the statistical difference, only the results for word n-grams for Spanish and bag of words for English and Spanish were different. From the table, we can see that TFIDF gave the highest accuracy for English tweets with 68.3%. The combination of the features gave the best results, 80.0%, for Spanish.

The results for age classification are given in Table 8.12. Among bag-of-words features, TFIDF gave the best result, while among word n-grams, word

**TABLE 8.11**

PAN 2015 Gender Classification Accuracy Results

| Feature | | English | Spanish |
| --- | --- | --- | --- |
| BOW | TF | 0.511 | 0.550 |
| | TFIDF | 0.683 | 0.690 |
| Text n-grams | Bigrams | 0.623 | 0.700 |
| | Trigrams | 0.652 | 0.520 |
| POS n-grams | Unigrams | 0.659 | 0.770 |
| | Bigrams | 0.659 | 0.750 |
| | Uni+bi | 0.660 | 0.780 |
| Combo | | 0.671 | 0.800 |

**TABLE 8.12**

PAN 2015 Age Classification Accuracy Results

| Feature | | English | Spanish |
|---|---|---|---|
| BOW | TF | 0.395 | 0.460 |
| | TFIDF | 0.691 | 0.470 |
| Word n-grams | Bigrams | 0.638 | 0.640 |
| | Trigrams | 0.553 | 0.460 |
| POS n-grams | Unigrams | 0.579 | 0.550 |
| | Bigrams | 0.611 | 0.670 |
| | Uni+bi | 0.593 | 0.700 |
| Combo | | 0.650 | 0.720 |

bigrams gave the best result. For POS n-grams, however, bigrams for English gave the best result while the combination of unigrams and bigrams gave the best result for Spanish. Looking at the statistical difference, only the results from bag-of-words features for English and word n-grams for Spanish were different from each other. From the table we can see that TFIDF features gave an accuracy of 69.1% for English; it is higher than the normalized term frequency and even the combination of the best features. For Spanish, however, the concatenation of the features gave the highest accuracy, 72.0%.

The average mean squared error across all five personality traits is given in Table 8.13. Most of the results are not statistically different from each other. Looking at the magnitude, TFIDF works best for English while the combination of all features works well Spanish. However, it should be noted that the magnitudes are marginal.

Our second experiment involved using word vector averages as features for classification run over a ten-fold cross-validation using the PAN 2016 dataset. The summary of the highest accuracy obtained for these experiments is given in Table 8.14. The highest accuracy obtained for English gender classification was 68.2% using a polynomial kernel with a degree of 3 and C = 100. There is more variety from these results given that the lowest

**TABLE 8.13**

PAN 2015 Average of Personality MSE Results

| Feature | | English | Spanish |
|---|---|---|---|
| BOW | TF | 0.029 | 0.033 |
| | TFIDF | 0.025 | 0.027 |
| Text n-grams | Bigrams | 0.027 | 0.029 |
| | Trigrams | 0.028 | 0.032 |
| POS n-grams | Unigrams | 0.035 | 0.126 |
| | Bigrams | 0.038 | 0.027 |
| | Uni+bi | 0.040 | 0.029 |
| Combo | | 0.029 | 0.026 |

**TABLE 8.14**

PAN 2015 English Gender
Classification Accuracy

| Task/Language | English | Spanish |
|---|---|---|
| Gender | 0.682 | 0.671 |
| Age | 0.448 | 0.513 |

was around 50.0%. Details of the results of other parameters are explicated in another paper in [38].

For age classification in English, the highest accuracy obtained was 44.8%. The SVM parameter that gave the best classification used a radial basis function kernel with C = 1 and gamma = 100, although most of the other values were close. The highest gender classification accuracy for Spanish tweets was 67.1%. This classifier used a radial basis function kernel with gamma = 1 and C = 100. On the other hand, the highest accuracy for Spanish age classification was 51.3%. This was given by a classifier that used a radial basis function kernel with gamma = 1 and C = 100.

Our third experiment uses CNNs to classify tweets from the PAN 2016 dataset using the hold-out setup. Table 8.15 shows the CNN accuracy results for English and Spanish.

There are many trends that can be gleaned from the previous table. First, we observe that using majority prediction for the tweets of a given dataset improves the result. This contrasts to other methods wherein the tweets for the user are concatenated. One exception would be gender classification for Spanish. Second, we note that increasing the dimensions increases the accuracy. Although the number of times when the accuracy improved is equal to that in which it worsened, the magnitudes tell otherwise. This is the same with changing the vectors to be trainable or non-trainable, non-static or static. Making the vectors trainable or non-static increases the accuracy. One explanation could be that the vectors get more finely tuned with more training data. There is an increase of 1.5% and 7.7% for English gender

**TABLE 8.15**

PAN 2016 Accuracy Comparison between Evaluation by Tweets and User

| | | English | | | | Spanish | | | |
|---|---|---|---|---|---|---|---|---|---|
| | | Age | | Gender | | Age | | Gender | |
| | | Static | Non-Static | Static | Non-Static | Static | Non-Static | Static | Non-Static |
| 100 | Tweet | 0.407 | 0.397 | 0.618 | 0.623 | 0.541 | 0.481 | 0.561 | 0.551 |
| | User | 0.481 | 0.473 | 0.651 | 0.667 | 0.547 | 0.560 | 0.557 | 0.550 |
| 300 | Tweet | 0.410 | 0.409 | 0.626 | 0.613 | 0.538 | 0.467 | 0.693 | 0.653 |
| | User | 0.496 | 0.473 | 0.643 | 0.721 | 0.547 | 0.547 | 0.507 | 0.587 |

classification using 100 and 300 dimensions respectively. There is also a 1.3% increase for Spanish age classification using 100 dimensions, while using 300 dimensions did not yield any difference. There is also an increase of 8.0% for Spanish gender classification using 300 dimensions. The other instances, however, lower the result when tuning the vectors. The biggest decrease is 4.0%, which comes from gender classification using 100 dimensions.

## 8.4 Conclusion and Future Work

After in-depth discussion of the two major aspects, aggression identification and author profiling, the following conclusion is derived.

In the case of aggression identification, plain/normal dense architecture outperforms other state-of-the-art architectures. The only downfall is that dense architecture could only be used in comparatively smaller datasets due to the indexing of each word encountered during modeling. The larger the index, the larger the matrix representation, resulting in higher usage of computational resources. Apart from that, words not seen in the dictionary/index are discarded. This could be tackled in future work. Moreover, the model does not capture the semantic meaning of the sentences like part of speech and morphology. This could also be considered in future work.

In the case of author profiling, looking over the PAN 2015 dataset, we can conclude that using a standard TFIDF can still outperform other features in terms of English classification tasks. A combination of various features, however, is better for the Spanish classification task. Meanwhile, regression over personality gives results that are close to each other. Using an average of word vectors as features for the PAN 2016 dataset for a classifier could have comparable results, although it is not as accurate since we are using different datasets. Finally, using a CNN architecture in conjunction with word vectors seems to yield a higher accuracy on average depending on the vector dimension size as well as vector trainability. However, we should have better tests for this in the future since the setups are currently different.

## References

1. Anna Schmidt, and Michael Wiegand. 2017. A Survey on Hate Speech Detection Using Natural Language Processing. In: Proceedings of the Fifth International Workshop on Natural Language Processing for Social Media, Association for Computational Linguistics, pages 1–10, Valencia, Spain.

2. Shervin Malmasi, and Marcos Zampieri. 2017. Detecting Hate Speech in Social Media. In: Proceedings of the International Conference Recent Advances in Natural Language Processing (RANLP), pages 467–472.

3. Irene Kwok, and Yuzhou Wang. 2013. Locate the Hate: Detecting Tweets Against Blacks. In: Proceedings of the Twenty-Seventh AAAI Conference on Artificial Intelligence, pages 1621–1622.

4. Chikashi Nobata, Joel Tetreault, Achint Thomas, Yashar Mehdad, and Yi Chang. 2016. Abusive Language Detection in Online User Content. In: Proceedings of the 25th International Conference on World Wide Web, pages 145–153., International World Wide Web Conferences Steering Committee.

5. Alexandra Schofield, and Thomas Davidson. 2017. Identifying Hate Speech in Social Media. XRDS: Crossroads, The ACM Magazine for Students, 24(2):56–59.

6. Björn Gambäck, and Utpal Kumar Sikdar. 2017. Using Convolutional Neural Networks to Classify Hate-speech. In: Proceedings of the First Workshop on Abusive Language Online, pages 85–90.

7. Ziqi Zhang, David Robinson, and Jonathan Tepper. 2018. Detecting Hate Speech on Twitter Using a Convolution-GRU Based Deep Neural Network. In: *Lecture Notes in Computer Science*, Springer Verlag.

8. Antigoni-Maria Founta, Constantinos Djouvas, Despoina Chatzakou, Ilias Leontiadis, Jeremy Blackburn, Gianluca Stringhini, Athena Vakali, Michael Sirivianos, and Nicolas Kourtellis. 2018. *Large Scale Crowdsourcing and Characterization of Twitter Abusive Behavior.* arXiv Preprint arXiv:1802.00393.

9. Karthik Dinakar, Roi Reichart, and Henry Lieberman, 2011. Modeling the detection of textual cyberbullying. In: *The Social Mobile Web*, pages 11–17.

10. Pete Burnap, and Matthew L Williams, 2015. Cyber hate speech on twitter: An application of machine classification and statistical modeling for policy and decision making. *Policy and Internet*, 7(2):223–242.

11. Ritesh Kumar, Aishwarya N Reganti, Akshit Bhatia, and Tushar Maheshwari. 2018b. Aggression-annotated Corpus of Hindi-English Code-mixed Data. In: Proceedings of the 11th Language Resources and Evaluation Conference (LREC), Miyazaki, Japan.

12. Ritesh Kumar, Atul Kr Ojha, Shervin Malmasi, and Marcos Zampieri. 2018a. Benchmarking Aggression Identification in Social Media. In: Proceedings of the First Workshop on Trolling, Aggression and Cyberbulling (TRAC), Santa Fe, USA.

13. Christos Baziotis, Nikos Pelekis, and Christos Doulkeridis. 2017. Datastories at semeval-2017 task 4: Deep lstm with attention for message-level and topic-based sentiment analysis. In: Proceedings of the 11th International Workshop on Semantic Evaluation (SemEval-2017), pages 747–754, *Vancouver, Canada, August.* Association for Computational Linguistics.

14. François Chollet et al., 2015. Keras. https://keras.io.

15. Martín Abadi, Paul Barham, Jianmin Chen, Zhifeng Chen, Andy Davis, Jeffrey Dean, Matthieu Devin, Sanjay Ghemawat, Geoffrey Irving, Michael Isard, Manjunath Kudlur, Josh Levenberg, Rajat Monga, Sherry Moore, G Murray, Benoit Steiner, Paul Tucker, Vijay Vasudevan, Pete Warden, Martin Wicke, Yuan Yu, and Xiaoqiang Zheng. 2016. Tensorflow: A system for large-scale machine learning. In: Proceedings of the 12th USENIX Conference on Operating Systems Design and Implementation, OSDI'16, pages 265–283, Berkeley, CA, USA: USENIX Association.

16. Jeffrey Pennington, Richard Socher, and Christopher D. Manning. 2014. Glove: Global vectors for word representation. In: *Empirical Methods in Natural Language Processing (EMNLP)*, pages 1532–1543.

17. Kashyap Raiyani, Teresa Gonçalves, Paulo Quaresma, and Vitor Beires Nogueira. 2018. "Fully Connected Neural Network with advance preprocessor to identify aggression over Facebook and twitter". In: Proceedings of the First Workshop on Trolling, Aggression and Cyberbullying (TRAC-2018). Santa Fe, New Mexico, USA: Association for Computational Linguistics, pages 28–41. URL: http://aclweb.org/anthology/W18-4404.

18. Efstathios Stamatatos. 2009. A survey of modern authorship attribution methods. *Journal of the American Society for Information Science and Technology*, 60(3):538–556.

19. Francisco Rangel, Fabio Celli, Paolo Rosso, Martin Potthast, Benno Stein, and Walter Daelemans. 2015. Overview of the 3rd Author Profiling Task at PAN 2015. In Linda Cappellato, Nicola Ferro, Gareth Jones, and Eric San Juan, editors, CLEF 2015 Evaluation Labs and Workshop – Working Notes Papers, 8-11 September, Toulouse, France. CEUR-WS.org, September 2015.

20. Francisco Rangel, Paolo Rosso, Ben Verhoeven, Walter Daelemans, Martin Potthast, and Benno Stein. 2016. Overview of the 4th Author Profiling Task at PAN 2016: Cross-genre Evaluations. In: Working Notes Papers of the CLEF 2016 Evaluation Labs, CEUR Workshop Proceedings. CLEF and CEUR-WS.org, September 2016.

21. Carlos E González-Gallardo, Azucena Montes, Gerardo Sierra, J Antonio Núñez-Juárez, Adolfo Jonathan Salinas-López, and Juan Ek. 2015. Tweets classification using corpus dependent tags, character and pos n-grams. In: CLEF (Working Notes).

22. Suraj Maharjan, and Thamar Solorio. 2015. Using wide range of features for author profiling. In: CLEF (Working Notes).

23. Mounica Arroju, Aftab Hassan, and Golnoosh Farnadi. 2015. Age, gender and personality recognition using tweets in a multilingual setting. In: 6th Conference and Labs of the Evaluation Forum (CLEF 2015): Experimental IR meets multilinguality, multimodality, and interaction, pages 23–31.

24. Fahad Najib, Waqas Arshad Cheema, and Rao Muhammad Adeel Nawab. 2015. Author's traits prediction on twitter data using content based approach. In: CLEF (Working Notes).

25. Scott Nowson, Julien Perez, Caroline Brun, Shachar Mirkin, and Claude Roux 2015. *Xrce Personal Language Analytics Engine for Multilingual Author Profiling*. Working Notes Papers of the CLEF.

26. Maite Giménez, Delia Irazú Hernández, and Ferran Pla. 2015. Segmenting target audiences: Automatic author profiling using tweets. In: Proceedings of CLEF.

27. Andreas Grivas, Anastasia Krithara, and George Giannakopoulos. 2015. Author profiling using stylometric and structural feature groupings. In: CLEF (Working Notes).

28. Alonso Palomino-Garibay, Adolfo T Camacho-Gonzalez, Ricardo A Fierro-Villaneda, Irazu Hernandez-Farias, Davide Buscaldi, Ivan V Meza-Ruiz, et al. 2015. A random forest approach for authorship profiling. In: CLEF (Working Notes).

29. Tomas Mikolov, Kai Chen, Greg Corrado, and Jeffrey Dean. 2013. Efficient Estimation of Word Representations in Vector Space. arXiv Preprint ArXiv:1301.3781.

30. Tomas Mikolov, Ilya Sutskever, Kai Chen, Greg S Corrado, and Jeff Dean. 2013. Distributed representations of words and phrases and their compositionality. In: *Advances in Neural Information Processing Systems*, pages 3111–3119.

31. Robert R McCrae, and Paul T Costa Jr. 1999. A five-factor theory of personality. Handbook of personality. *Theory and Research*, 2:139–153.

32. Helmut Schmid 2013. Probabilistic part-ofispeech tagging using decision trees. In: New Methods in Language Processing, Page, 154.

33. Radim Řehůřek, and Petr Sojka. 2010. Software Framework for Topic Modelling with Large Corpora. In: Proceedings of the LREC 2010 Workshop on New Challenges for NLP Frameworks, pages 45–50, Valletta, Malta. ELRA. http://is.muni.cz/publication/884893/en.

34. Corinna Cortes, and Vladimir Vapnik. 1995. Support-vector networks. *Machine Learning*, 20(3):273–297.

35. Yoon Kim. 2014. Convolutional neural networks for sentence classification. *Arxiv Preprint Arxiv*:1408.5882.

36. RF Woolson 2007. Wilcoxon Signed-Rank Test. Wiley Encyclopedia of Clinical Trials, pages 1–3.

37. Roy Bayot, and Teresa Gonçalves 2016. *Multilingual Author Profiling Using Svms and Linguistic Features*.

38. Roy Bayot, and Teresa Gonçalves. 2016. Author Profiling Using Svms and Word Embedding Averages—Notebook for Pan at clef 2016.

.

# 9

## The Effect of Using Features Computed from Generated Offline Images for Online Bangla Handwritten Character Recognition

Shibaprasad Sen, Ankan Bhattacharyya and Kaushik Roy

**CONTENTS**

## 9.1 Introduction

The primary aim of online handwriting recognition (OHR) systems is to provide the textual (Unicode) interpretation of a given sequence of handwritten strokes. In general, a stroke is represented by a sequence of pen points (x, y, t), where (x, y) is the position of the pen point and a timestamp t. Though research on OHR started in the late 1990s, researchers are currently showing renewed interest in this domain due to following reasons:

(a) Widespread usage of mobile devices (many of which are equipped with touchscreens), which, due to their affordability and easy availability, have become part of life. These devices can record online handwritten input easily [1].

(b) The intention to support texts written in regional scripts for which no standard keyboard layout is commonly adopted (like Khmer) and which are difficult to type on a soft keyboard.

(c) The advancement of research in areas like deep learning, speech recognition, optical character recognition and machine translation, which can be adapted to the task of OHR.

In this chapter, we have focused on the generation of a system to recognize handwritten Bangla characters online. The reason for working with Bangla is not only the limited presence of research in this area, but also the strong cultural heritage of the language, which motivates us to do some substantial work to take advantage of cutting-edge machine learning approaches with the required customization. Besides, Bangla is the second most popular language in India. It is one of the official languages and scripts used in India and it is the official language of Bangladesh. The script is inherited from the ancient Brahmi script. The writing style of Bangla is from left to right like other Indo–Aryan scripts, such as Devanagari. The basic character set of the Bangla script contains 39 consonants and 11 vowels, which are shown in Figure 9.1 [2].

In the work described in this chapter, we aim to build an offline character image database from the corresponding online information and at the same time to generate a more powerful feature set to increase overall character recognition accuracy by incorporating both the online and offline properties of

| অ | আ | ই | ঈ | উ |
| ঊ | এ | ঐ | ও | ঔ |
| ক | খ | গ | ঘ | ঙ |
| চ | ছ | জ | ঝ | এ |
| ট | ঠ | ড | ঢ | ণ |
| ত | থ | দ | ধ | ন |
| প | ফ | ব | ভ | ম |
| য | র | ল | শ | ষ |
| স | হ | ড় | ঢ় | য় |
| ৎ | ং | ঃ | ঁ | |

**FIGURE 9.1**
Bangla alphabet set.

the character images. In this chapter, we have mentioned the complete recognition setup, from the preparation of an online handwritten Bangla character database, to pre-processing, to the details of feature extraction from pre-processed data and finally to the recognition of the handwritten characters by some popularly used classifiers [3,4]. This chapter also suggests some ways to enhance the strength of the features generated by online information by integrating corresponding offline information into it [5]. Though the first commercial OHR systems were available in the 1990s, their abilities were limited and success rates not so good. We believe that many areas of OHR are still not solved, and here we briefly summarize some of the challenges faced by different researchers in the recognition of online handwritten Bangla characters:

i) A wide variation in the writing styles of different individuals makes the recognition procedure challenging.

ii) There is extensive writing speed and size variation even for the same person. Faster writing speed means that fewer pixel points are used to write the same information [6].

iii) The variation in the number of strokes [7] used to write a character introduces challenges for the recognition of online handwritten characters. This is illustrated in Figure 9.2, where the character is written using two, three and a maximum of 4 strokes, respectively.

iv) Variation in the order of the strokes used to write a character makes the thing tougher. For example, in Figure 9.3 four strokes are used to write the character 'ঝ' but the occurrences of strokes are different.

v) Shape similarity between some characters poses another challenge, as can be seen in Figure 9.4.

vi) Different users have different writing styles. Figure 9.5 illustrates a scenario where the same character 'ক' is written in different styles.

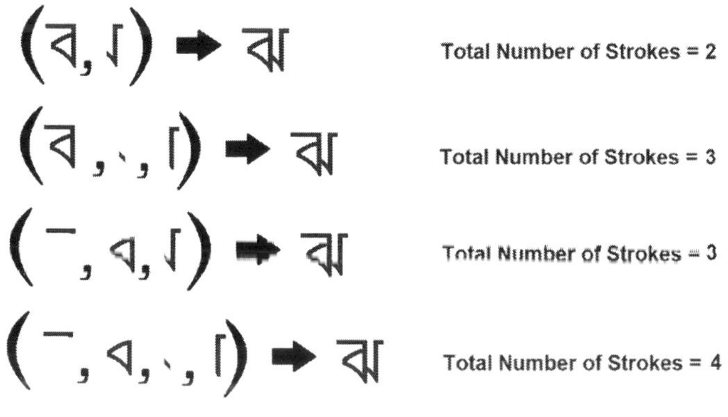

**FIGURE 9.2**
Illustration of writing the same character 'ঝ' using different number of strokes.

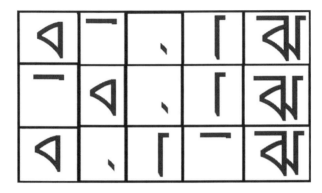

**FIGURE 9.3**
Stroke order variation to write the same character.

**FIGURE 9.4**
Example of some similarly shaped characters.

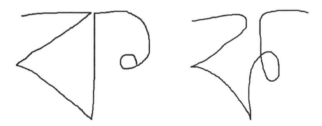

**FIGURE 9.5**
Illustration of different writing styles to write the same character.

Due to these various challenges, the increasing popularity of online devices and the desire to convert written text into editable electronic content, researchers in OHR are paying more attention to online handwritten character recognition for different languages [8]. Some research materials on Devanagari [9–16], English [17–24], and Gurumukhi/Gujarati [25–37] scripts are available in literature. A few works involve new-age machine learning concepts like deep learning [38–45] and extreme learning machines [46]. Regarding Bangla script, while some research publications do exist (discussed in the following), progress is still at in infant stage.

Roy et al. in [2] proposed the holistic features extraction technique, which collects both sequential and dynamic information from pen movements on writing pads. These features are evaluated by a quadratic classifier to recognize online handwritten basic Bangla characters. The authors in [47] have shown the effectiveness of direction code and point-float histogram features for the recognition of four popular Indic scripts. The authors used some popular classifiers like the multi-layer perceptron (MLP), nearest neighbor (NN) and hidden Markov model (HMM) for classification. A different scheme is highlighted in [48], where the constituent strokes are first extracted at the character level. The previously mentioned sequential and dynamic features are extracted from strokes for their recognition. A recognized stroke sequence is then matched with an already built rule base for the construction of characters.

Parui et al in [49] highlighted a stroke-based character recognition mechanism where one HMM is built for each stroke class. The authors manually grouped the strokes of all characters into 54 classes depending on the similarity of shapes at the grapheme level. Characters are formed from recognized stroke sequences with the help of a previously built look-up table. Direction code-based features were used by authors to recognize online handwritten Bangla basic characters in [50]. Another stroke-based character recognition approach was mentioned by Bandyopadhyay et al. in [51], where they generated shape-based features from constituent strokes collected from all the character samples. The recognition of strokes was performed by using the dynamic time warping (DTW) technique [52]. The author in [53] also experimented with a stroke-based character recognition scheme where he generated features by dividing the scheme into nine zones and considered inclination, the direction of the writing, curliness, standard deviation of x and y coordinates and curvature of the strokes. An SVM classifier [54] was used for stroke classification. The effectiveness of using distance-based features, global features, local features, hausdorff distance-based features, area features, mass distribution and chord length features for online handwritten Bangla characters was discussed in detail by Sen et al. in [55–58].

These research works have made efforts towards the enhancement of recognition accuracy of online handwritten Bangla characters by introducing new features and combining different feature vectors [59]. In this chapter, an effort has been made to use the information of offline images with features generated in online mode for the enhancement of overall character recognition accuracy. The purpose of this work is two-fold: First, the generation of offline character images from online information that may be used for working offline, and second, the enhancement of character recognition accuracy by incorporating the offline information of its online counterparts.

The organization of the paper is as follows. Section 9.2 introduces a brief review of some of the well-known feature extraction techniques for the recognition of online handwritten Bangla characters. Section 9.3 deals with database preparation and the pre-processing mechanism. Section 9.4 describes the feature generation procedure used in both online and offline modes to recognize the online handwritten Bangla characters. The classification results of different classifiers for online and offline modes (individually) are reported in Section 9.5. This section also specifies the mechanism for including offline information with features produced in online mode to increase recognition accuracy. Finally, Section 9.6 concludes the experiment procedure and describes future planning.

## 9.2 Literature Review

### 9.2.1 Direction Code-Based Feature [50]

The collected character data firstly go through some pre-processing stages before this feature extraction technique is applied. The pre-processing steps involve elimination of redundant (repeated) pen points, and then the resultant points are re-sampled into a fixed number of points that are almost equidistant (Figure 9.6).

Each of the pre-processed character samples is segmented into K sub-divisions and the direction code features are estimated for all the sub-divisions. As the character sample may contain single or multiple strokes, the number of subdivisions for the $i$th stroke is calculated by Equation 9.1.

$$K_i = \text{round}\left(\frac{X_i}{X} K\right) \qquad (9.1)$$

The length of strokes $X_i$ is computed by summing the distance between the two consecutive points in the sequence that compose it. The total length $X$ of the character sample is calculated as $X = \sum_i X_i$ and the value of $i$ denotes the number of strokes used to represent sample character sample. If $\sum_i K_i \neq K$ then a slight modification needs to make $\sum_i K_i = K$.

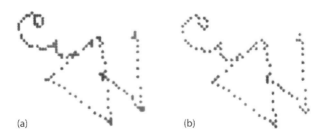

(a)                                                      (b)

**FIGURE 9.6**
(a) Pen points of a character having three component strokes. (b) The pen points of the character after first re-sampling.

However, if the number of pen points of the $i$th stroke is not a multiple of $K_i$, then another re-sampling is needed to generate a new sequence of points, such that the number becomes the nearest multiple of $K_i$ and the points are almost equidistant.

The sub-division of strokes is illustrated in Figure 9.7. Considering that the character sample has three strokes, those strokes contain 46, 27 and 28 sample pen points after the pre-processing stage. The computed ratio of the lengths of these three strokes is 4:3:3. So, if 10 subdivisions are considered for the character sample, then the three strokes will have 4, 3, and 3 subdivisions respectively. Now it is observed that the number of points of the first stroke is not a multiple of 4. So, the authors have re-sampled the first stroke again and this time the number of points in the first stroke should be 48, which is the nearest multiple of 4. Similarly, the third should have 27 points after the second re-sampling, which is the nearest multiple of 3, whereas the second one would not undergo further re-sampling as 27 is already a multiple of 3. Figure 9.7b and 9.7c demonstrate the second re-sampling.

After dividing the character sample into $K$ segments, the direction code is measured. Let's consider that the $j$th stroke contains the sequence of pen points $p_1, p_2, ..., p_{ni}$. Again, consider that the angle made with the x-axis when moving from position $p_b$ to $p_{b+1}$ be $\alpha_b$, $b = 1, 2, ..., ni - 1$ ( $0° \leq \alpha_b < 360°$). In this feature extraction approach, the information describing the change in direction when traversing from one pen point to the next plays an important role. The direction change from one pen point to the next can be denoted by following Freeman's direction code (eight possible values from 1, 2, ..., 8) [60].

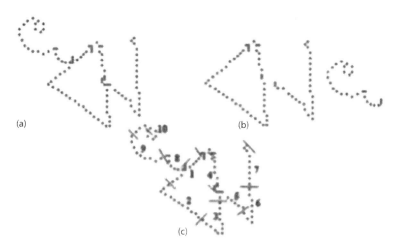

(a)

(b)

(c)

**FIGURE 9.7**
(a) Effect of second time re-sampling for the character shown in Figure 9.6a. (b) Three component strokes of the character according to the writing order. (c) Number of subdivisions made for each stroke.

The direction code for the first pixel of a stroke is considered to be 0. Hence, each stroke of the character is represented by direction code measurements.

A histogram corresponding to measured direction codes is computed for each subdivision. As the measurement of changes in directions is mentioned by one of the nine possible values (from 0 to 9), the histogram contains nine components for each subdivision. If the whole character contains $K$ number of subdivisions then this procedure produces a $9 \times N$ element feature vector.

### 9.2.2 Area and Local Features

Research by Sen et al. [58] shows where they find the strength of some features (by computing area, mass distribution and chord length) in recognizing online handwritten Bangla basic characters. The features are extracted from the local regions of the character by firstly dividing the sample into a quad-tree-based image segmentation approach mentioned in [58]. The detailed feature extraction mechanisms are described in the following.

#### 9.2.2.1 Area Feature

In most cases, the shapes of Bangla characters are different from each other. Hence, the structure of constituent parts of different characters in different blocks must be different when viewed at various quad-tree depths. Keeping this fact in mind, we have assumed that block-wise computation of the area under the curve of such character segments may generate discriminative features which in turn may play an important role towards the enhancement of character recognition accuracy. The block-wise computation of area under the curve is illustrated in Figure 9.8. There may exist certain blocks with no pixel point and in this case the computed area for this zone is considered to be 0. Looking into the structural patterns of the blocks for a sample like অ, it can be stated that the produced feature values may play a useful role in classifying characters efficiently. The composite Simpson's rule, shown in Equation 9.2, has been used to find the zone-wise area under the curve.

$$\text{Area} = \sum_{k=1}^{n-1} \left( \left( \frac{h}{3} \right) * \left( y_k + 4 * y_{middle} + y_{k+1} \right) \right) \tag{9.2}$$

Where $n$ denotes the number of pen points in each block from $y_1$ to $y_n$. ($y_k$, $y_{k+1}$) are the $y$ coordinate values of the consecutive pen points in that block. The measurement of $h$ and $y_{middle}$ are computed by evaluating Equations 9.3 and 9.4.

$$h = \frac{\text{mod}\left(x_{k+1} - x_k\right)}{2} \tag{9.3}$$

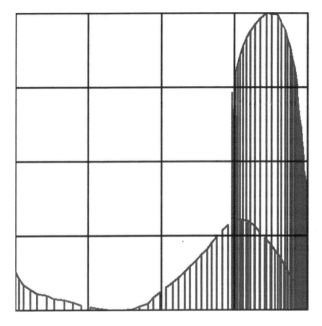

**FIGURE 9.8**
Illustration of block-wise area feature computation for two character samples using composite Simpson's rule at quad-tree depth two.

$$y_{\text{middle}} = \frac{\left(y_k + y_{k+1}\right)}{2} \tag{9.4}$$

For a particular character sample, a total of $N*N$ number of values are estimated by evaluating Equation 9.1 while the sample character is segmented into $N \times N$ blocks. These computed area values are considered features.

### 9.2.2.2 Local Feature

#### 9.2.2.2.1 Mass Distribution

Mass distribution describes the number of data points within a block at different quad-tree depths. As the shapes of Bangla characters are different, the numbers of pixels in different blocks are different. Hence, a block-wise count of the pixels may play a vital role in distinguishing differently shaped characters. The measured mass distribution for the sample আ is reflected in Figure 9.9 at quad-tree depth 3. The gray pixels denote pixels in the blocks.

#### 9.2.2.2.2 Chord Length

Chord length features are estimated by computing the length of the character segments in each block and considered features to differentiate different

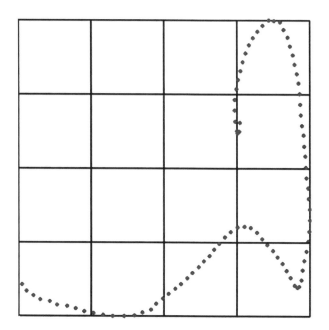

**FIGURE 9.9**
Illustration of mass distribution feature computation at quad-tree depth.

character shapes. If $p_1$, $p_2$, $p_3$, ..., $p_n$ is the sequence of pixel points for a certain block, then the length of this segment is measured by summing the distances between consecutive points from $p_1$ to $p_n$. As the structural patterns of different characters are different, it is found that the computed length of the segments varies significantly. Thus, the produced feature vector may be effective for this pattern classification problem. The detailed procedure for the computation of this feature is mentioned in [58].

### 9.2.3 Point-Based Feature [2, 48]

In this feature extraction technique, the character sample is represented by the sequence $t = [t_1,...,t_N]$ of feature vectors $t_i = (t_{i1}, t_{i2}, t_{i3})^\mathsf{T}$.

Here, $t_{i1}$, $t_{i2}$ denote the pen co-ordinates normalized by the sample mean $\mu$ and measured as:

$$\mu = \frac{1}{N}\sum_{i=1}^{N}p_i$$

$$t_{i1} = \frac{(x_i - \mu_x)}{\sigma_y}$$

$$t_{i2} = \frac{(y_i - \mu_i)}{\sigma_y}$$

The standard deviation $(\sigma_y)$ is computed as:

$$\sigma_y = \sqrt{\frac{1}{N-1} \sum_{i=1}^{N} (\mu_y - y_i)^2}$$

and $t_{i3}$ is measured as follows:

$t_{i3} = \arg((x_{i+1} - x_{i-1}) + j*(y_{i+1} - y_{i-1}))$, with $j^2 = -1$ and "arg" the phase of the complex number above is an approximation of the tangent slope angle at point $i$.

### 9.2.4 Transition Count Feature [61]

The proposed technique firstly scales a sample character into a window of size $N \times N$ and then the scaled sample is binarized. The value of the position is set to 1 if this is a character data point (referred to as a foreground pixel), and set to 0 otherwise (background pixel). The center of gravity (CG) of the binarized image is computed and then the character image is segmented into four sub-regions. The transition counts from foreground to background pixels and vice-versa are measured along four directions, row, column and along two major diagonals (i.e. at four different angles $\theta = 0°, 90°, 45°, 135°$) for each sub-region. This technique basically finds data distribution along four directions. The left side of Figure 9.10 shows the binarized character image divided into four sub-regions based on CG. It has also been noted that the maximum value of the transition count for a particular row/column/diagonal within a sub-image varies from 0 to 6. In the next step, the frequencies of these counts are estimated and act as features. For each sub-image, 28 (i.e. 7 × 4) feature values are generated for 0 to 1 transition and another 28 for 1 to 0. Therefore, a total of 56 (i.e. 28 × 2) feature values are generated for each sub-image. Finally, a total of 224 (i.e. 56 × 4) feature values are produced for the whole image. For easy reference, feature calculation (here, 0 to 1 transition) on a hypothetical image is described on the right side of Figure 9.10.

### 9.2.5 Topological Feature [61]

#### 9.2.5.1 Crossing Point

This feature is useful for differentiating a few character pairs in the Bangla alphabet that are strongly similar in shape. For example, 'ঝ' and 'ঝ' are quite structurally similar, except 'ঝ' has a crossing point on the lower left side due to the formation of a loop. Therefore, the presence of the crossing point plays an important role in distinguishing such character pairs. To estimate this feature, the pixel points of the character sample are re-sampled in such way that the points are at unit distance apart. This feature examines if a character sample contains a loop or not. The pixel points of the sample character are

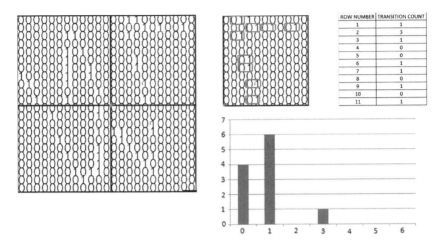

| ROW NUMBER | TRANSITION COUNT |
|---|---|
| 1 | 1 |
| 2 | 3 |
| 3 | 1 |
| 4 | 0 |
| 5 | 0 |
| 6 | 1 |
| 7 | 1 |
| 8 | 0 |
| 9 | 1 |
| 10 | 0 |
| 11 | 1 |

**FIGURE 9.10.**
Transition count feature calculation for character 'অ'.

treated as crossing points that have more than three adjacent neighbors to help to form a loop. For example, consider Figure 9.11, where the existence of a loop can easily be observed; in this figure, the crossing points are marked by light grey circles, while the dark grey squares represent the neighbors of the corresponding crossing point. A total of five feature values (the number of crossing points, the coordinates of the first crossing point and the coordinates of closest neighbor of the first crossing point) are measured by using this technique. This procedure enhances recognition accuracy by distinguishing similar shapes based on the existence of loops.

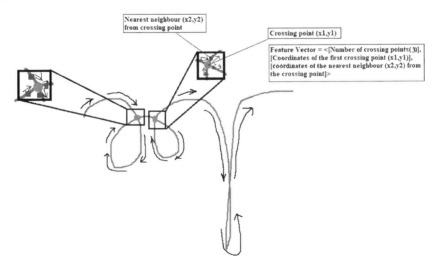

**FIGURE 9.11**
Identification of a crossing point in character sample 'শ'.

## 9.3 Database Preparation and Pre-processing

### 9.3.1 Design of the Data Collection Form

The primary stage of the database development process is the design of a data collection form or forms. During the design of this form, various factors are considered. Different factors in the populace may influence handwriting traits, like gender, age, educational qualification, religion, etc. Additionally, writing at different intervals can affect the handwriting style of different individuals. To capture these variations and to simplify the problem, data collection forms are designed that contain characters of Bangla script that are mostly used for writing. Isolated character data collection forms and handwritten text data collection forms are designed separately. Isolated characters include all the basic characters of the alphabet available in Bangla along with numerals and vowel modifiers. The data collection form is divided into two zones: In the header zone writer information is collected, and in the writing zone writers need to write the suggested characters in their own handwriting. The left side of Figure 9.12 shows a blank data collection form of isolated characters where all the fields are marked for easy understanding, whereas the right side of the figure shows the corresponding filled-in form. During

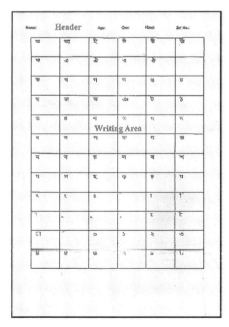
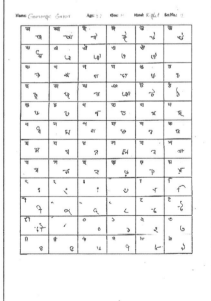

**FIGURE 9.12**
One blank character data collection form (left) and one filled-in form (right).

the design of the form, all the vowels and alphabets that are used in writing Bangla texts are considered along with vowel modifiers. A total of 11 vowels, 40 consonants, 10 vowel modifiers and 10 numerals are considered in the design of the isolated character data collection form.

During data collection, individuals are asked to write the sample isolated characters and sample printed text within the provided area on the form at their natural speed. The writers are also given each data collection form (blank) on different days, so that intra-writer variations along with inter-writer variations can be captured. Individuals of different gender, culture, age groups and educational qualifications are considered during data collection. Individuals from 8 to 70 years old are considered during data collection, including schoolchildren, retired persons, graduates, postgraduate students and people with different professions. The natural variations are incorporated into the database to create standard isolated character and text document databases [62,63] to be helpful in assessing whether a character/text recognition algorithm is robust in nature or not. For a character-level database, characters are written within the boxes as shown in Figure 9.12. In the next step, the coordinates of the boundary of every box are measured and then the character-level information ($x,y$ coordinates, along with pen status) from all the boxes are stored.

This section describes the detailed statistics of two different types of databases developed here. One hundred volunteers contributed during the design of the isolated character database, and from each individual three sets of characters of the Bangla alphabet have been collected. Though initially 135 volunteers agreed to contribute to the development of this character database, only 100 volunteers completed the required three sets. The database contains approximately 12,000 alphabets, 3,300 vowels, 3,000 vowel modifiers and 3,000 numerals. Each collected dataset contains ordered pixel points with $x$ and $y$ coordinates, along with pen up/down information; this is saved in .dat format with a particular naming convention: < 4 *digit serial number (Writer_id)* >_< 2 *digit serial number (Set_id)* >. For example, the first isolated digitized form would be named '0000_01', where '0000' stands for the ID of the first writer, and '01' for the first set of writings by that writer; these two fields are separated by the "_" sign.

## 9.4 Feature Extraction

### 9.4.1 Directed Hausdorff Distance (DHD)-Based Features

An existing HD-based feature has been extracted in the present work from handwritten characters using online information and from the offline images generated from online information.

The popular maxmin function has been considered to compute forward HD h(P,Q) from set P to set Q by using Equation 9.5.

$$h(P,Q) = \frac{\max}{x \in P}\left\{\frac{\min}{y \in Q}\left\{d(x,y)\right\}\right\} \qquad (9.5)$$

Where pen points of sets P and Q are represented by $a$ and $b$, respectively. For the present experiment, d(x,y) measures the Euclidian distances between x and y. In the same way, h(Q,P) (known as backward HD) has been estimated from set Q to set P. Figure 9.13a illustrates the calculation of distances from each point of zone P to every other point of zone Q and the minimum valued distances for each point of zone P are marked in yellow. Among those minimum valued distances, the maximum is considered as forward HD and reflected in Figure 9.13b. Similarly, the computation of backward HD is shown in Figure 9.14a and b.

During analysis it has been noticed that the computed h(P,Q) and h(Q,P) values are not the same most of the time. Therefore, in this work, the authors have measured both the forward and backward HD values (known as DHD) from each zone to all other zones to get exclusivity. This feature extraction strategy generates $N$–1 DHD values for each zone and thereby a total of $N*(N$–1) DHD features are estimated for an entire character.

To extract DHD-based features in offline mode, the previously generated character images are binarized and at quad-tree depth two, and the previously mentioned procedure used in online mode is repeated by considering object pixels

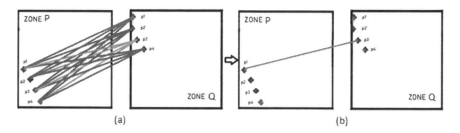

**FIGURE 9.13**
(a,b) Illustration of forward HD computation.

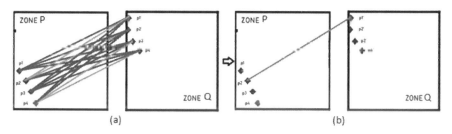

**FIGURE 9.14**
(a,b) Illustration of backward HD computation.

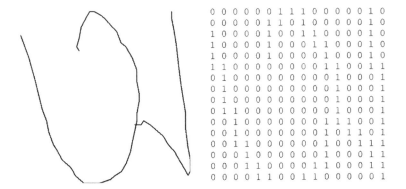

**FIGURE 9.15**
Character image generated from online information and corresponding binarized image.

of the image. The character image generated from online information and the corresponding binarized image are shown in Figure 9.15.

## 9.5 Experimental Results and Analysis

The feature set extracted through the DHD procedure from both online and offline modes of the character images is evaluated by using four well-known classifiers, multi-layered perceptron (MLP), support vector machine (SVM), BayesNet and simple logistic, where the character sample is divided into 16 zones (at quad-tree depth two). A five-fold cross-validation scheme has been used for classification purposes. Table 9.1 records the classification accuracy produced by the classifiers mentioned for both online and offline modes of the character images.

Table 9.1 demonstrates that the strength of the feature vector produced using online information is higher than the feature set generated from offline character images for all the classifiers.

**TABLE 9.1**

Recognition Accuracies of Different Classifiers for DHD-Based Feature Used in Both Online and Offline Modes of Character Images

| Classifiers | DHD | | Online + Offline |
| --- | --- | --- | --- |
| | **Online Mode** | **Offline Mode** | |
| MLP | 95.57 | 93.54 | 95.92 (with pca 92.61) |
| BayesNet | 88.61 | 84.99 | 88.37 |
| SVM | 94.43 | 91.8 | 93.78 |
| Simple Logistic | 95.36 | 92.72 | 94.89 |

**TABLE 9.2**

Some Misclassified Character Pairs

| Original Character (200 samples) | Misclassified as (Samples out of 200) |
|---|---|
| ভ | উ(8) |
| ৱ | ল(৪) |
| খ | ঘ(6) |
| ঙ | উ(4) উ(4) |
| ঘ | য(4) ম(4) |

Feature vectors produced in online and offline modes are also combined to observe the effect of the combination. At this stage, MLP produces the best accuracy, 95.92%, which is higher than the accuracies obtained when the features are extracted either from online or offline mode of the character images. The obtained result ensures that the strength of the feature vector produced in online mode increases by combining offline information into it. At this stage, a few misrecognized characters are shown in Table 9.2.

When we minutely observed the prediction scores of the classifier MLP for both the feature vectors, it was observed that for some cases the prediction score of a correct choice was higher for the offline feature vector. We have also observed that the prediction score of a correct choice was not on top for both the features. Hence, in order to increase the overall character recognition performance, a late fusion technique was tried [30] on the top five prediction choices by MLP, i.e. after classification of the characters by both the feature vectors (online mode and offline mode) using MLP. In this scheme, the MLP classifier was trained for each modality ($m$), "online mode" and "offline mode". Each modality was assigned an expertise trust score $e_{sm}$ that is specific to each activity $s$ (prediction choice). This score is defined as a normalized performance computed from the performance measure $P_{sm}$ (observed probability/prediction percentage from classifier) obtained by the modality $m$ for the activity $s$ as defined in Equation 9.6.

$$e_{sm} = \frac{P_{sm}}{\sum_m P_{sm}} \tag{9.6}$$

The final decision is computed as follows: for each observation, a label is allocated to the activity with the best trusted expert score of both modalities.

For each observation $O_i$, we have applied the argmax fusion operator as defined in Eq. 9.7.

$$s_i = \text{argmax}\left(e_{s_i \text{ online based}}, e_{s_i \text{ offline based}}\right) \tag{9.7}$$

The application of the late fusion technique with a max fusion operator gives the character prediction choice with the highest trusted score between both modalities. Hence, the resultant choice may be selected from either

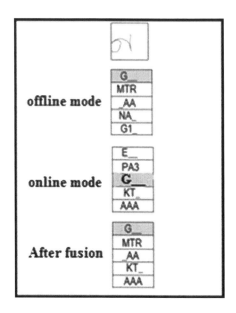

**FIGURE 9.16**
Illustration of prediction choices of a character symbol using online, offline and late fusion approaches.

point-based or curvature-based modality. As a result, overall stroke recognition accuracy increases, reaching 96.74%.

Figure 9.16 illustrates the outcome of applying the late fusion technique after classification by MLP. This figure highlights the top-five recognition choice of 'গ' (G__) by MLP for both the feature vectors generated by online and offline mode. By observing the figure, it can be stated that features from offline mode recognize the character correctly as the top-most choice. On the other hand, the features generated in online mode fail to recognize the character as the top choice. When late fusion is applied, based on the classifier confidences, the character sample is then recognized correctly, which in turn increases the overall system performance.

After applying the late fusion technique, the count of misclassified characters reduces, as is reflected in Table 9.3.

**TABLE 9.3**

Misclassified Characters after Adopting Late Fusion Technique

| Original Character (200 samples) | Misclassified as (Samples out of 200) |
| --- | --- |
| ভ | ড(5) |
| ন | ল(4) |
| খ | ঘ(5) |
| ঙ | উ(2) ড(1) |
| ঘ | য(3) ম(2) |

## 9.6 Conclusion

In the present work, the efficiency of the feature set generated from online handwritten Bangla character images is increased by creating a corresponding offline version of the same. The intention is not only to enhance the character recognition accuracy but also to build both online and offline character databases that can be used for further work in this domain. Though the combined feature set reflects better performance than the individual feature vectors, which were separately created from online and offline versions of the character images, the overall accuracy is not much increased. The applied late fusion scheme produces better results and improves recognition accuracy. However, due to strong structural similarity, a few characters are still misclassified. Hence, we plan to introduce more features to differentiate these similarly shaped characters in future, in order to improve the overall recognition accuracy. Another viable plan is to test the efficiency of this feature set for online handwritten texts written in other languages.

## References

1. S. Agarwal, and V. Kumar. 2005. Online character recognition. In: Proceedings of the 3rd International Conference on Information Technology and Applications, 698–703.
2. F. Alvaro, and R. Zanibbi. 2013. A shape-based layout descriptor for classifying spatial relationships in handwritten math. In: Proceedings of the ACM Symposium on Document Engineering, 123–126.
3. C. Bahlmann, and H. Burkhardt. 2004. The writer independent online handwriting recognition system frog on hand and cluster generative statistical dynamic time warping. *IEEE Trans. Pattern Anal. Mach. Intell.* 26(3), 299–310.
4. M. Farha, G. Srinivasa, A. J. Ashwini, and K. Hemant. 2013. Online handwritten character recognition. *Int. J. Comput. Sci.* 11(5), 30–36.
5. C. C. Tappert, C. Y. Suen, and T. Wakahara. 1990. The state of online handwriting recognition. Transaction on Pattern Analysis and Machine Intelligence. *IEEE* 12(8), 787–808.
6. G. D. Vescovo, and A. Rizzi. 2007. Online handwriting recognition by the symbolic histograms approach. In: Proceedings of the International Conference on Granular Computing. IEEE, 686–690.
7. M. F. Zafar, D. Mohammad, and M. M. Anwar. 2006. Recognition of online isolated handwritten characters by back propagation neural nets using sub-character primitive features. In: Proceedings of the 9th International Conference on INMIC. IEEE, 157–162.

8. W. Zhao, J. F. Liu, and X. L. Tang. 2002. Online handwritten English word recognition based on cascade connection of character HMMs. In: Proceedings of the International Conference on Machine Learning and Cybernetics. IEEE, 1758–1761.

9. R. K. Bawa, and R. Rani. 2011. A preprocessing technique for recognition of online handwritten gurmukhi numerals. In: Proceedings of the International Conference on High Performance Architecture and Grid Computing. IEEE, 275–281.

10. M. Gupta, N. Gupta, and R. Agrawal. 2012. Recognition of online handwritten Gurmukhi strokes using support vector machine. In: Proceedings of the 7th International Conference on Bio-Inspired Computing: Theories and Application, 495–506.

11. J. R. Prasad, and U. Kulkarni. 2015. Gujarati character recognition using adaptive neuro fuzzy classifier with fuzzy hedges. *Int. J. Mach. Learn. Cybernet.* 6(5), 763–775.

12. M. K. Sachan, G. Lehal Singh, and V. K. Jain. 2011. A novel method to segment online Gurmukhi script. In: Proceedings of the International Conference on Information Systems for Indian Languages, 1–8.

13. M. K. Sachan, G. Lehal Singh, and V. K. Jain. 2011. A system for online Gurmukhi script recognition. In: Proceedings of the International Conference on Information Systems for Indian Languages, 299–300.

14. K. C. Santosh, and J. Iwata. 2012. Stroke-based cursive character recognition. In: Advances in Character Recognition, X. Ding, ed. INTECH Open.

15. K. C. Santosh, C. Nattee, and B. Lamiroy. 2012. Relative positioning of stroke-based clustering: A new approach to online handwritten Devanagari character recognition. *Int. J. Image Graph.* 12(2), 2 (2012), 25 pages. Doi: 10.1142/S0219467812500167.

16. K. C. Santosh, and L. Wendlingc. 2015. Character recognition based on non-linear multi-projection profiles measure. *Int. J. Front. Comput. Sci.* 9(5), 678–690.

17. A. Sharma, and K. Dahiya. 2012. Online handwriting recognition of Gurmukhi and Devanagiri characters in mobile phone devices. In: Proceedings of the International Conference on Recent Advances and Future Trends in Information Technology, 37–41.

18. A. Sharma, R. Kumar, and R. K. Sharma. 2008. Online handwritten Gurmukhi character recognition using elastic matching. In: Proceedings of the Congress on Image and Signal Processing. IEEE, 391–396.

19. A. Sharma, R. Kumar, and R. K. Sharma. 2009. Rearrangement of recognized strokes in online handwritten Gurmukhi words recognition. In: Proceedings of the 10th International Conference on Document Analysis and Recognition, 1241–1245.

20. A. Sharma, R. Kumar, and R. K. Sharma. 2010. HMM-based online handwritten Gurmukhi character recognition. *Int. J. Mach. Graph. Vision.* 19(4), 439–449.

21. P. K. Singh, R. Sarkar, N. Das, S. Basu, and M. Nasipuri. 2014. Statistical comparison of classifiers for script identification from multi-script handwritten documents. *Int. J. Appl. Pattern Recogn.* 1(2), 152–172.

22. B. B. Ahmed, S. Naz, M. I. Razzak, S. F. Rashid, M. Z. Afzal, and T. M. Breuel. 2016. Evaluation of cursive and noncursive scripts using recurrent neural networks. *Int. J. Neur. Comput. Appl.* 27(3), 603–613.

23. M. R. Bouguelia, S. Nowaczyk, A. Verikas, and K. C. Santosh. 2017. Agreeing to disagree: Active learning with noisy labels without crowdsourcing. *Int. J. Mach. Learn. Cybernet.*, 1–13.

24. J. Du, J. Zhai, and J. Hu. 2017.Writer adaptation via deeply learned features for online Chinese handwriting recognition. *Int. J. Doc. Anal. Recogn.* 20(1), 69–78.

25. A. El-Sawy, H. EL-Bakry, and M. Loey. 2016. CNN for handwritten Arabic digits recognition based on LeNet-5. In: Proceedings of the International Conference on Advanced Intelligent Systems and Informatics, 566–575.

26. J. D. Prusa, and T. M. Khoshgoftaar. 2017. Improving deep neural network design with new text data representations. *Int. J. Big Data* 4(1), 7 (2017). Doi: 10.1186/s40537-017-0065-8.

27. X. Tang, Y. Ding, and K. Hao. 2017. A novel method based on line-segment visualizations for hyper-parameter optimization in deep networks. *Int. J. Pattern Recogn. Artif. Intell.* 32(3) (2017), 1851002-1–1851002-15.

28. W. Wang, G. Tan, and H. Wang. 2017. Cross-domain comparison of algorithm performance in extracting aspect-based opinions from Chinese online reviews. *Int. J. Machine. Learn. Cybernet.* 8(3), 1053–1070.

29. Q. Wu, Z. Gui, S. Li, and J. Ou. 2017. Directly connected convolutional neural networks. *Int. J. Pattern Recogn. Artif. Intell.* 32(5), 1859007-1–1859007-17.

30. A. Fu, C. Dong, and L. Wang. 2015. An experimental study on stability and generalization of extreme learning machines. *Int. J. Mach. Learn. Cybernet.* 6(1), 129–135.

31. K. Roy, N. Sharma, and U. Pal. 2007. Online Bangla handwriting recognition system. In: Proceedings of the International Conference on Advances in Pattern Recognition, 117–122.

32. T. Mondal, U. Bhattacharya, S. K. Parui, K. Das, and D. Mandalapu. 2010. Online handwriting recognition of Indian scripts - The first benchmark. In: Proceedings of the 12th International Conference on Frontiers in Handwriting Recognition, 200–205.

33. K. Roy. 2012. Stroke-database design for online handwriting recognition in Bangla. *Int. J. Mod. Eng. Res.* 2(4), 2534–2540.

34. S. K. Parui, K. Guin, U. Bhattacharya, and B. B. Chaudhuri. 2008. Online handwritten Bangla character recognition using HMM. In: Proceedings of the International Conference on Pattern Recognition, 1–4.

35. U. Bhattacharya, B. K.Gupta, and S. K. Parui. 2007. Direction code based features for recognition of online handwritten characters of Bangla. In: Proceedings of the International Conference on Document Analysis and Recognition, 58–62.

36. A. Bandyopadhyay, and B. Chakraborty. 2009. Development of online handwriting recognition system: A case study with handwritten Bangla character. In: Proceedings of the World Congress on Nature and Biologically Inspired Computing, 514–519.

37. R. Ghosh. 2015. A novel feature extraction approach for online Bengali and Devanagari character recognition. In: Proceedings of the International Conference on Signal Processing and Integrated Networks, 483–488.

38. S. Sen, R. Sarkar, and K. Roy. 2016. A simple and effective technique for online handwritten Bangla character recognition. In: Proceedings of the 4th International Conference on Frontiers in Intelligent Computing: Theory and Application, 201–209.

39. N. Joshi, G. Sita, A. G. Ramakrishnan, and V. Deepu. 2005. Machine recognition of online handwritten Devanagari characters. In: Proceedings of the International Conference on Document Analysis and Recognition, 1156–1160.

40. S. Sen, A. Bhattacharyya, A. Das, R. Sarkar, and K. Roy. 2016. Design of novel feature vector for recognition of online handwritten Bangla basic characters. In: Proceedings of the 1st International Conference on First International Conference on Intelligent Computing & Communication, 485–494.

41. S. Sen, R. Sarkar, K. Roy, and N. Hori. 2016. Recognize online handwritten Bangla characters using hausdorff distance based feature. In: Proceedings of the 5th International Conference on Frontiers in Intelligent Computing: Theory and Application, 541–549.

42. S. Sen, M. Mitra, S. Chowdhury, R. Sarkar, and K. Roy. 2016. Quad-tree based image segmentation and feature extraction to recognize online handwritten Bangla characters. In: Proceedings of the 7th IAPR TC3Workshop on Artificial Neural Networks in Pattern Recognition. Ulm, Germany, 246–256.

43. J. Demsar. 2006. Statistical comparisons of classifiers over multiple data sets. *Int. J. Mach. Learn. Res.* 7, 1–30.

44. M. Hall, E. Frank, G. Holmes, B. Pfahringer, P. Reutemann, and I. H. Witten. 2009. The WEKA data mining software: An update. *ACM SIGKDD Explor. Newslett.* 11(1), 10–18.

45. J. Pinquier, S. Karaman, L. Letoupin, P. Guyot, R. Megret, J. Benois-Pineau, Y. Gaestel, and J. Dartigues. 2012. Strategies for multiple feature fusion with Hierarchical HMM: Application to activity recognition from wearable audiovisual sensors. In: International Conference on Pattern Recognition, 3192–3195.

46. S. Sen, A. Bhattacharyya, P. K. Singh, R. Sarkar, K. Roy, and D. Doermann. 2018. Application of structural and topological features to recognize online handwritten Bangla characters. *ACM Trans. Asian Low-Resour. Lang. Inf. Process.*, 17(3), Doi: 10.1145/3178457.

47. K. C. Santosh, C. Nattee, and B. Lamiroy. 2012. Relative positioning of stroke-based new approach to online handwritten Devanagari character recognition. In: *Int. J. Image Graph.*, 25 pages.

48. K. C. Santosh. 2011. Character recognition based on DTW-Radon*ICDAR*, Doi: 10.1109/ICDAR.2011.61.

49. K. C. Santosh, C. Nattee, and B. Lamiroy. 2010. Spatial similarity based stroke number and order free clustering. In: 12th International Conference on Frontiers in Handwriting Recognition, 652–657.

50. S. Kubatur, M. Sid-Ahmed, and M. Ahmadi. 2012. A neural network approach to online Devanagari handwritten character recognition. In: Proceedings of the International Conference on High Performance Computing and Simulation. Doi: 10.1109/HPCSim.2012.6266913

51. A. Kumar, and S. Bhattacharya. 2010. Online Devanagari isolated character recognition for the iPhone using hidden Markov models. In: Proceedings of the International Conference on Students' Technology Symposium, 300–304.

52. H. Freeman. 1974. Computer processing of line-drawing images. *ACM Comput. Surv.* 6(1), 57–97.

53. V. L. Lajish, and S. K. Kopparapu. 2014. Online handwritten Devanagari stroke recognition using extended directional features. In: Proceedings of the 8th International Conference on Signal Processing and Communication System. IEEE. Doi: 10.1109/ICSPCS.2014.7021063.

54. Sk. Md. Obaidullah, C. Halder, K. C. Santosh, N. Das, and K. Roy. 2017. PHDIndic_11: Page-level handwritten document image dataset of 11 official Indic scripts for script identification. *Multimedia Tool. Appl.*, Doi: 10.1007/s11042-017-4373-y.

55. Sk. Md. Obaidullah, K. C. Santosh, C. Halder, N. Das, and K. Roy. 2017. Word-level multi-script Indic document image dataset and baseline results on script identification. *Int. J. Comput. Vis. Image Process.*, 7(2), 81–94.

56. A. Tripathi, S. S. Paul, and V. K. Pandey. 2012. Standardisation of stroke order for online isolated Devanagari character recognition for iPhone. In: Proceedings of the International Conference on Technology Enhanced Education. IEEE, 1–5.

57. Gerrit J. J. van den Burg, and Patrick J. F. Groenen. 2016. GenSVM: A generalized multiclass support vector machine. *J. Mach. Learn. Res.* 17(225), 1–42.

58. H. Swethalakshmi, C. C. Sekhar, and V. S. Chakravarthy. 2007. Spatiostructural features for recognition of online handwritten characters in Devanagari and Tamil scripts. In: Proceedings of the International Conference on Artificial Neural Networks, 230–239.

59. H. Swethalakshmi, A. Jayaraman, V. S. Chakravarthy, and C. C. Sekhar. 2006. Online handwritten character recognition for Devanagari and Telugu scripts using support vector machines. In: Proceedings of the International Workshop on Frontiers in Handwriting Recognition, 367–372.

60. Sk. Md. Obaidullah, C. Halder, N. Das, and K. Roy. 2016. A new dataset of word-level offline handwritten numeral images from four official Indic scripts and its benchmarking using image transform fusion. *Int. J. Intell. Eng. Inf.*, 4(1), 1–20.

61. Sk. Md. Obaidullah, K. C. Santosh, N. Das, C. Halder, and K. Roy. 2018. Handwritten Indic script identification in multi-script document images: A survey. *Int. J. Pattern Recognit. Artif. Intell.* 32(10), 1856012.

62. K. C. Santosh, and L. Wendling. 2015. Character recognition based on non-linear multi-projection profiles measure. *Front. Comput. Sci.*, 9(5), 678–690.

63. S. D. Connell, R. M. K. Sinha, and A. K. Jain. 2000. Recognition of unconstrained online Devanagari characters. In: Proceedings of the 15th International Conference on Pattern Recognition, 368–371.

# 10

## Handwritten Character Recognition for Palm-Leaf Manuscripts

Papangkorn Inkeaw, Jeerayut Chaijaruwanich
and Jakramate Bootkrajang

**CONTENTS**

## 10.1 Introduction

Optical handwritten character recognition (OHCR) is a well-known automatic method for transforming document images of handwriting to machine-encoded texts. Although several techniques of OHCR have been developed during the past decade, OHCR is still a growing research field. Furthermore, OHCR techniques are nowadays applied to office automation systems. Also, recent research directions are broadening the use of OHCR techniques in the processing of ancient documents in digital libraries. Challenges in OHCR are often associated with types of writing media and the complexity of alphabets. Each writing medium requires a specific preprocessing technique, while the characteristics of the alphabet dictate the design of character segmentation and recognition methods.

Performing OHCR on ancient palm-leaf manuscripts poses considerable difficulties. The archived palm-leaf manuscripts are often in poor condition and thus careful preprocessing steps are needed to deal with various kind of noises. The absence of line markers also adds some uncertainties during line segmentation. More importantly, the majority of palm-leaf manuscripts were widespread among South Asian and Southeast Asian languages, where the combination of possibly touching and broken alphabets, vowels and tone marks complicates the matter even further.

In this chapter, we shall elucidate the problems associated with performing OHCR on palm-leaf manuscripts. We then outline current state-of-the-art preprocessing, line segmentation, character segmentation and character recognition steps which were successfully applied to the task. We finally conclude the chapter with a working example of an ancient Lanna Dhamma character recognition system.

## 10.2 Palm-Leaf Manuscripts

Palm-leaf manuscripts are ancient documents made out of dried palm leaves. Palm leaves were primarily used as writing materials in South Asia and Southeast Asia in ancient times. Over hundreds of years, the manuscripts have become damaged. In an attempt to preserve these potentially valuable documents, several organizations have started surveying local communities everywhere. A lot of invaluable manuscripts have been recovered, especially in Sri Lanka, India, Laos, Burma, Cambodia and Thailand. Figure 10.1 shows examples of palm-leaf manuscripts, found in the northern part of Thailand, describing traditional medicine which is important for further investigation. These manuscripts were preserved and archived using microfilming, scanning and photographing techniques. Normally, several palm leaves were put into a single image called a folio. Storing palm-leaf manuscripts in folio form helps reduce digitization time and storage space.

## 10.3 Challenges in OHCR for Palm-Leaf Manuscripts

Storing palm-leaf manuscripts in the form of imagery data somewhat limits interaction with the documents. Instead, it is more desirable to transform

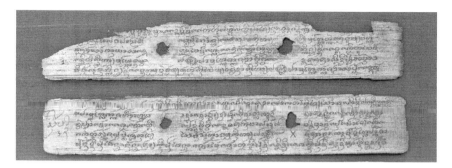

**FIGURE 10.1**
Examples of palm-leaf manuscripts found in Northern Thailand.

the palm-leaf images into machine-encoded texts so that the document can be edited, retrieved, accessed and processed conveniently. Unfortunately, it is rather cost-prohibitive to perform image-to-text transformation manually. OHCR is a viable option for addressing the problem. Generally, there are four tasks in the OHCR process, namely preprocessing, segmentation, recognition and postprocessing. Challenges in OHCR are often associated with the types of writing media and the complexity of alphabets. Here we shall elucidate the common issues associated with performing OHCR on palm-leaf manuscripts in each of the four steps. These problems should be addressed during the development of an OHCR system for palm-leaf manuscripts.

The quality of the images is often the first issue encountered in the OHCR process. Many palm-leaf manuscript images are taken without standardized equipment, yielding wild differences in image quality. Although digitizing using a professional grade scanner is usually preferred, curators often opt for digital cameras when it comes to digitizing highly valuable and fragile originals. Thus, it is difficult to ensure that the subjects sit perfectly flat. Nonuniform illumination also leads to poor contrast between the text and the background. In addition, most of the palm-leaf images were taken without any line markers on the background. Hence, there exists some visible skewness, which is the rotation of a document image from the horizontal axis, in the resulting images.

Several kinds of noises and image artifacts on palm-leaf manuscript images further worsen the situation. According to the study in [1], noise on palm-leaf manuscripts can be categorized into three types, namely small component noise, leaf streak and leaf hole. Examples of these noises are illustrated in Figure 10.2. Small component noises appear as grains on old document images. They may be generated not only by the natural degradation of writing media but also by the image binarization process. Furthermore, the manuscripts are dried palm leaves, naturally prepared without the knowledge of modern paper-making processes, and as a result leaf streaks are still visible on the writing media. The leaf streak noises usually appear along the cracks of the leaf. Lastly, to bind the manuscripts, palm leaves were pierced

**FIGURE 10.2**
Three types of noises presented on palm-leaf manuscripts.

with one or two holes around the area in the middle of the page. The holes become artifacts that should be removed before further processing. Due to these challenges, effective image processing techniques are required in the preprocessing step of an OHCR system.

For palm-leaf manuscripts, each individual character is carved using a sharp writing tool on long, rectangular-cut dried palm leaves. One of the difficulties in extracting text lines from palm-leaf manuscripts is the fact that the lines are cramped up together with very little spacing between them. In addition, text lines may also be slanted or curved upward/downward on account of being handwritten. Furthermore, unlike the Latin alphabets where characters are sequenced from left to right, the alphabets mostly found in palm-leaf manuscripts were written in a multi-level style. Some characters are positioned below or above another character. They tend to go far out of their main line, touching or overlapping with other characters from adjacent lines. One needs to be aware of these issues when trying to perform line segmentation as well as character segmentation on palm-leaf manuscript folios.

Lastly, the huge number of character classes, variations in handwriting styles, unconstrained writing and existence of visually similar characters are considered to be challenging issues in character recognition tasks. Moreover, as most alphabets appearing on palm-leaf manuscripts are ancient characters which are not in use in modern days, only a handful of labeled examples from a limited number of writers are provided for training a recognition model. In OHCR from palm-leaf manuscripts, the lack of training examples becomes a major challenge in constructing an effective recognition model.

## 10.4 Document Processing and Recognition for Palm-Leaf Manuscripts

Document image processing and recognition have received considerable attention from the research community for their potential in many applications and their technical challenges. Many effective methods for document image preprocessing, segmentation, recognition and postprocessing have been proposed to date.

## 10.4.1 Preprocessing

In the preprocessing task, an input document image is transformed into a suitable format, hopefully removing most of the noise as well as undergoing foreground and background separation, where the pixels corresponding to the characters (considered the foreground) and the pixels corresponding to the background are commonly designated with binary 1s and 0s, respectively. The preprocessing task for palm-leaf manuscripts additionally consists of binarization and skew correction [1, 2].

### 10.4.1.1 Binarization

Binarization is a common starting point for document image analysis that converts gray image values into binary representation. The performance of binarization techniques has a great impact and directly affects the performance of the segmentation and recognition tasks. A simple and well-known binarization approach is the thresholding method. Based on the choice of the thresholding value, binarization methods can generally be divided into two types, global thresholding and local adaptive thresholding. For global thresholding, a single threshold value is calculated from the global characteristics of the image. Otsu's method [3], for example, is a well-known global binarization technique. In local adaptive thresholding, the threshold value is calculated in each smaller local image area, region or window. Many local adaptive thresholding techniques have been proposed, for example the adaptive mean method, Niblack's method [4], Sauvola's method [5], Wolf's method [6] and the NICK method [7]. Another binarization approach is the training-based method. This approach uses classifiers to classify each pixel into either the foreground or background pixel. The top two training-based binarization methods were proposed in the Binarization Challenge for the ICFHR 2016 Competition on the Analysis of Handwritten Text in Images of Balinese Palm Leaf Manuscripts [8]. The best method in this competition, ICFHR G1 [8], employs a fully convolutional network (FCN). It takes a color sub-image as input and outputs the probability that each pixel in the sub-image is part of the foreground. The second-best method, ICFHR G2 [8], uses two neural network classifiers to classify each pixel as background or not. A comparative study of document image analysis methods for palm-leaf manuscripts is reported in [9]. The study performed a comprehensive test of binarization, text line segmentation, isolated character recognition, word recognition and transliteration with collections of palm-leaf manuscripts from Southeast Asia (i.e. Balinese, Khmer and Sundanese palm-leaf manuscripts). It reported that Niblack's method, ICFHR G1 and ICFHR G2 performed well for Sundanese, Khmer and Balinese manuscripts, respectively, with the highest F-measure scores. A comparison of thresholding methods shows that Niblack's and NICK methods achieved the highest F-measure score for Balinese and Khmer manuscripts, respectively.

### 10.4.1.2 Noise Reduction

After the document image is transformed into a binary image, unwanted components that are considered noise are to be removed. A noise reduction technique aimed at removing the three types of noises mentioned in Section 10.2 is proposed in [1]. The small component noises were filtered out by analyzing the size of the connected components, e.g. a group of touching foreground pixels, while the leaf streaks were detected by examining the ratio of the width to the height of each component. To deal with the leaf hole, a mask was generated by applying global thresholding to the input image, and then it was used to remove the leaf holes.

### 10.4.1.3 Skew Correction

To correct document skewness, the angle at which a document image is rotated as referenced by the horizontal axis, the orientation of the input document image should first be determined. There are several commonly used methods for detecting document skewness. Some methods perform line detection using the Hough transform [10, 11]. The median of the angles of the detected lines is taken to be the document skewness. Another technique relies on the detection of connected components [12, 13], which are roughly equivalent to characters, and then finds the average angles connecting their centroids. In contrast, some methods straightforwardly detect document skewness using a projection profile analysis [14, 15]. A document page is projected at several angles, and the variances in the number of foreground pixels per projected line are computed. The projection parallel to the correct alignment of the lines will likely have the maximum variance. After the document skewness has been detected, the document image must be rotated with the angle in order to correct this skewness.

## 10.4.2 Segmentation

In the segmentation task, a binary document image is gradually separated to obtain several isolated characters or word images as the results. It involves line segmentation and character segmentation.

### 10.4.2.1 Text Line Segmentation

Most document analysis and recognitions start with line segmentation. A variety of projection profile techniques are used in [1, 16–19]. The goal of those methods is to extract estimated medial or seam positions of text lines based on the peaks or valleys of the horizontal projection profile of the document image. Another text line segmentation approach tries to find optimal paths that pass through text line seams [20–23]. These line segmentation methods operate on a binary document image, and so an effective initial binarization method is required. By contrast, some line segmentation methods that do not rely on the effectiveness of the binarization process have also been proposed. A method proposed in [24] created the medial axis of the text lines by analyzing the horizontal projection profiles. Two adjacent lines are separated by the seam-carving algorithm. An adaptive path-finding method has been proposed in [25]. A competitive algorithm is presented to find the medial axis of the text lines. To deal with a component spreading over multiple lines, two adjacent lines are separated by using a modified A* path-planning technique [26]. A comparative study [9] showed that the adaptive path-finding method achieved better results than the seam-carving method on three palm-leaf manuscript datasets.

### 10.4.2.2 Character Segmentation

After line segmentation, characters in each text line are to be segmented into isolated character images which will be recognized in the recognition task. Character segmentation methods can be categorized into three strategies, namely dissection, recognition-based segmentation and holistic methods [27]. The dissection strategy is considered a heuristic approach. Since different alphabets have different characteristics, the dissection method is rather specific to the alphabet it was designed for. The characteristics which are normally used are contour information [28, 29], reservoir [30] and skeleton [29, 31]. Recognition-based segmentation, on the other hand, utilizes feedback from a character recognition model in order to determine segmentation points. A recent study showed that recognition-based methods generally outperform dissection methods [32]. The recognition-based segmentation methods usually consist of two phases, namely an over-segmentation phase and a segmentation point-searching phase. Normally, a dissection method is used in the over-segmentation phase to obtain candidate segmentation points. The candidate segmentation points are used to segment a given word image into several fragments. The fragments are represented in an appropriate data structure that represents all possible segmentation hypotheses. Most recent methods for touching digits usually represent possible segmentation hypotheses in a directed acyclic graph (DAG) [33–35]. The problem is then how to search for the best segmentation path. Existing approaches have considered using the Viterbi algorithm [33, 36] and the shortest path algorithm [34, 37]. Alternatively, an undirected graph can be used to represent segmentation hypotheses [36]. The best segmentation hypothesis is a set of linear subgraphs obtained from applying a modified Viterbi algorithm. Despite the current availability of various recognition-based character segmentation methods, these existing methods cannot satisfyingly segment characters belonging to some complex writing styles. A recognition-based segmentation method, designed for tackling a rather complex handwritten character, is probably the one for handwritten Thai characters [38]. The method analyzes the reservoir properties of a word image to obtain candidate segmentation points. Optimal segmentation points are determined using a greedy algorithm. In our previous work [39], candidate segmentation points are obtained by analyzing background and foreground skeletons. These points are used to obtain the redundant fragments of a word image. The fragments are then used to form a segmentation hypotheses graph. The hypotheses graph is partitioned into subgraphs, each corresponding to a segmented character using the partitioning algorithm. Unlike the previous two approaches, the holistic segmentation methods avoid the need to segment a word image into meaningful fragments. It seeks to recognize a word image in one go. For that purpose, a labeled dataset of word images is required to train the holistic segmentation model. We will discuss this approach as word recognition in the next section.

### 10.4.3  Recognition

Common to the majority of OHCR pipelines is the character recognition task, where an image actually gets converted into text; it is the task where existing methods have been unable to provide satisfactory results. Character recognition methods can be categorized into two streams, namely segmentation-based and segmentation-free methods.

#### *10.4.3.1 Segmentation-Based Approach*

Segmentation-based methods require isolated character images as input. An appropriate feature extraction and effective classifier can increase recognition accuracy. Many feature extraction methods for handwritten character recognition have been proposed in the literature. Notable feature extraction methods are the histogram of orientation gradient (HOG) [40], scale invariant feature transform descriptor [41], image moment descriptors [42–48], Fourier descriptors [49], multi-projection profiles [50, 51] and chain codes [49, 52, 53]. Furthermore, some researchers have reported that combining several feature descriptors to form a feature vector can improve recognition performance [54–56]. To recognize isolated character images, there are several classification models available, for example k-nearest neighbor (kNN), support vector machines (SVM) and multi-layer perceptron (MLP), to name just a few.

Another character recognition approach is to adopt convolutional neural networks (CNN) as the recognition model. CNNs perform feature extraction simultaneously with learning the recognition model. A well-known CNN architecture is probably the LeNet-5 [57], which performs reasonably well on handwritten digit classification. Several architectures for character recognition have been presented to date, such as SPnet [58], Vanilla [9] and SHL-CNN

[59]. Normally, a CNN requires a large training sample to perform at its best due to the classifier's high complexity. In our experience, educating a CNN model with a small training set could lead to serious overfitting.

### 10.4.3.2 Segmentation-Free Approach

In segmentation-free character recognition, a word image can be sent directly as an input to the classifier. A popular recognition model in this approach is the combination of a recurrent neural network-long short-term memory (RNN-LSTM) [60] and a connectionist temporal classification (CTC) [61]. RNNs are similar to the feed-forward neural networks with the exception that they can use the internal state (memory) to process sequences of inputs. However, the vanishing gradient problem, where the gradients of the weight vectors become too small to be useful for learning, is often encountered during the training phase. LSTM architecture was then introduced to deal with the vanishing gradient problem. The LSTM network adds multiplicative gates and additive feedback to the RNN [60]. Recent studies on word recognition using the RNN-LSTM can be found in [62, 63].

## 10.5 An OHCR System for Lanna Palm-Leaf Manuscripts

In this section, we shall present a working OHCR system for Lanna palm-leaf manuscripts as an example. These manuscripts are important sources of the knowledge of Lanna wisdom. The system consists of four main tasks, as shown in Figure 10.3, namely preprocessing, layout analysis, segmentation and character recognition. In the preprocessing task, palm-leaf images are segmented from a folio image. Each palm-leaf image is enhanced and transformed into a binary image. Next, in the layout analysis task, text layouts in the binary image are detected. Each text layout is individually decomposed into isolated character images in the segmentation task. Finally, in the character recognition task, all the characters are recognized by a classifier. A machine-encoded text of the input palm-leaf image is produced as the output.

### 10.5.1 Preprocessing

The system starts the OHCR process with the preprocessing task. As palm-leaf manuscripts are usually archived in the form of folio images containing multiple palm-leaf images, each palm leaf needs to be extracted from the folio image in order to facilitate the process of OHCR. To do that, a page segmentation method for palm-leaf manuscripts has been proposed [64]. The method consists of two main steps: (1) object detection and (2) object

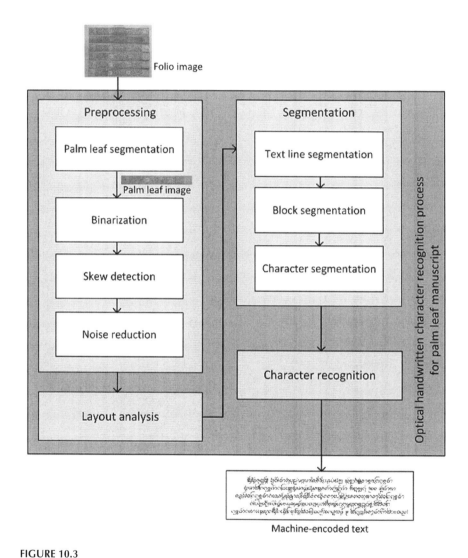

**FIGURE 10.3**
Flowchart of optical handwritten character recognition process for Lanna palm-leaf manuscripts.

selection. In the first step, a folio image in RGB color space is transformed into L*a*b* color space. All pixels represented by a* and b* are clustered into k groups. Then, a labeled image is generated by using a cluster number as a label of each pixel. Finally, each connected component is represented as an object in the folio image. The second step is a rule-based selection of objects as either palm leaf or not palm leaf. The three rules were designed based on the shape information of palm-leaf manuscripts. After a palm-leaf image is extracted from a folio image, it is transformed into a binary image using the NICK method. Prior to noise reduction, skew correction is performed by taking advantage of leaf streak noises because leaf streaks are usually parallel with text lines on a palm-leaf page. The angle of leaf streaks to the x-axis can be used to infer the skewness of the image. The document skewness is detected by using the Hough transform. The next step is the noise reduction step. To deal with leaf hole noises, the background plate and palm leaf are distinguished. The original palm-leaf image in RGB color is transformed into L*a*b* color. All pixels represented by a* and b* are clustered into two groups. Assuming that the palm leaf is the largest object in the original image, we can intuitively assign the largest cluster to the group of palm-leaf pixels. Next, a labeled image is generated using a cluster number as a label of each pixel. Since leaf hole noises are usually a part of palm leaf and touch the background, components in the binary image that are adjacent to background pixels are detected. Those components are removed from the binary image. For leaf streak noises, the binary image is smoothed by filters, as shown in Figure 10.4. Furthermore, the connected components that are similar to a straight line and parallel with the x-axis are removed from the binary image. Lastly, small component noises are removed by considering connected components with extremely small size.

## 10.5.2 Layout Analysis

By observing several Lanna palm-leaf manuscripts, we found four patterns of layout alignment as illustrated in Figure 10.5. For these patterns, the text layouts are arranged from the left to the right of a palm leaf. They can be separated by some vertical lines. To find these separating lines, the vertical projection profile of the binary palm-leaf image is obtained and analyzed. We assumed that gaps between characters in the same layout are obviously smaller than the gap between layouts. Before computing the projection profile, we need to reduce the sensitivity caused by gaps among characters in order to merge characters in the same layout into one component. Hence, the morphology dilation [65] operation is applied to the binary image. Then, the vertical projection profile is computed. Next, all pairs of opening and closing points of hills are detected. Each pair corresponds to each layout on the given binary palm-leaf image.

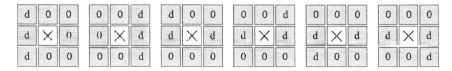

**FIGURE 10.4**
Six filters that used for smoothing a binary palm leaf image. The pixel in the center of the filter is the target. Pixels overlaid by any square masked "d" are ignored. If the pixels overlaid by the square masked "0" all have a value of zero then the value of the target pixel is forced to have the same value.

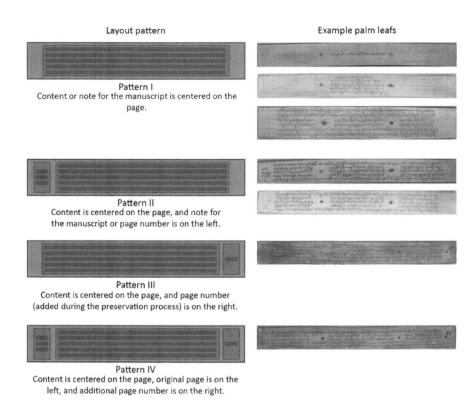

Layout pattern                              Example palm leafs

**Pattern I**
Content or note for the manuscript is centered on the page.

**Pattern II**
Content is centered on the page, and note for the manuscript or page number is on the left.

**Pattern III**
Content is centered on the page, and page number (added during the preservation process) is on the right.

**Pattern IV**
Content is centered on the page, original page is on the left, and additional page number is on the right.

**FIGURE 10.5**
Four patterns of text layouts found on Lanna Dhamma palm-leaf manuscripts.

### 10.5.3 Segmentation

The next task in the OHCR process is the segmentation step, which greatly affects the recognition performance. A binary image of each text layout is gradually separated to obtain several isolated character images. We divided this task into three steps, namely (1) text line segmentation, (2) block segmentation and (3) character segmentation.

In text line segmentation, we adopted the concepts of the text line segmentation method presented in [18]. The number of text lines in the given binary image is firstly determined. Unlike [18], the number of text lines is estimated by analyzing the horizontal projection profile of the whole image. The number of peaks in the projection profile is assigned to be the number of text lines, because the overall curvature of lines tends to be linear, though some sections may be curved.

In block segmentation, text lines are separated into small blocks, disregarding their semantics. This task is required in order to prepare suitable input images for the character segmentation, which works better when the input is a small block of characters. It can be easily done by analyzing a vertical projection profile of each text line. Firstly, the vertical projection profile of each text line is computed. Gaps between characters are then extracted from the projection profile by scanning to find the positions where the profile is zero. To avoid cutting a broken character into redundant blocks, the smaller gaps are ignored. Meanwhile, the larger gaps are used as cutting points to separate the text line into several small blocks.

In character segmentation, characters in each block are separated into isolated characters. To do that, we proposed a recognition-based character segmentation method in [39]. The method consists of three phases: (1) determination of candidate segmentation points, (2) representation of segmentation in a hypotheses graph and (3) determination of the best segmentation hypothesis. In the first phase, we analyzed the skeletons of a given word image to obtain a set of candidate segmentation points. The candidate segmentation points are used to separate the word image into fragments. In the second phase, we represented all segmentation hypotheses as an undirected graph. In the last phase, the graph was partitioned into subgraphs corresponding to segmented character images. We defined a confidence function that combines feedback from a recognition model together with the expected width and height of segmented characters to guide the partitioning algorithm. A multilayer perceptron was used as the recognition model. We designed a dynamic programming algorithm for finding the global optimal solution, in the sense that it maximizes the confidence function, to the graph partitioning problem.

### 10.5.4 Recognition

Character recognition is the last task of our OHCR system. In this task, each isolated character is to be classified into its corresponding character class. To enhance the performance of writer-independent recognition accuracy, we proposed an efficient character recognition method of handwritten Lanna Dhamma characters using intelligent moment feature selection (LDIMS) [54]. The LDIMS obtains a set of optimally designed features from a comprehensive set of shape-based moment descriptors. It consists of three main phases: (1) determination of moment orders for each effective moment descriptors, (2) determination of the best combination of moment descriptors and (3) optimized moment feature selection using an inheritable bi-objective combinatorial genetic algorithm. The moment features are considered because they provide scale and rotation invariants as well as noise robustness. Through the experiments, four effective moment descriptors including Meixner, Charlier, Tchebichef and Hahn were selected to form a feature vector. A kNN was employed as the classifier as it makes less assumption on the distribution of the data.

## 10.6 Concluding Remarks and Outlook

In this chapter, we have discussed the issues associated with performing OHCR on palm-leaf manuscripts. Those problems should be emphasized during the development of OHCR systems for palm-leaf manuscripts. We then outlined the current state-of-the-art preprocessing, line segmentation, character segmentation and character recognition steps which were successfully applied to the task. Finally, a working example of an ancient Lanna Dhamma character recognition system was presented. Despite the availability of the various methods proposed to date, there is still room to improve on OHCR systems for palm-leaf manuscripts. Compared to the human recognition ability and the requirement of applications, the current recognition accuracy by OHCR systems for palm-leaf manuscripts is still insufficient. Furthermore, most recent methods in OHCR were proposed with a deep learning approach, but this cannot be adopted in the works for palm-leaf manuscripts because of the limitation of the training dataset. Constructing large training datasets is needed to develop efficient OHCR methods for palm-leaf manuscripts that use the deep learning approach.

## References

1. Inkeaw, P., et al., Lanna Dharma handwritten character recognition on palm leaves manuscript based on Wavelet transform. In: 2015 IEEE International Conference on Signal and Image Processing Applications (ICSIPA). 2015.
2. Klomsae, A., Image feature extraction for Lanna Dharma handwritten character recognition, In: Department of Computer Science. Chiang Mai University, Thailand. 2012.
3. Otsu, N., A threshold selection method from gray-level histograms. *IEEE Transactions on Systems, Man, and Cybernetics*, 1979. 9(1): p. 62–66.
4. Niblack, W., *An Introduction to Digital Image Processing*. Strandberg Publishing Company. p. 215. 1985.
5. Sauvola, J., and M. Pietikäinen, Adaptive document image binarization. *Pattern Recognition*, 2000. 33(2): p. 225–236.
6. Wolf, C., J. Jolion, and F. Chassaing, Text localization, enhancement and binarization in multimedia documents. *Object Recognition Supported by User Interaction for Service Robots*, 2002. 2: p. 1037–1040.
7. Khurshid, K., et al., Comparison of Niblack inspired binarization methods for ancient documents. In: IS&T/SPIE Electronic Imaging. SPIE, 2009.
8. Burie, J.-C., et al., ICFHR2016 competition on the analysis of handwritten text in images of Balinese palm leaf manuscripts. 2016 [cited 2018 6/12/2018]; Available from: http://amadi.univ-lr.fr/ICFHR2016_Contest/.

9. Kesiman, W.M., et al., Benchmarking of document image analysis tasks for palm leaf manuscripts from Southeast Asia. *Journal of Imaging*, 2018. 4(2): p. 43.

10. Amin, A., and S. Fischer, A document skew detection method using the Hough transform. *Pattern Analysis and Applications*, 2000. 3(3): p. 243–253.

11. Gari, A., et al., Skew detection and correction based on Hough transform and Harris corners. In: International Conference on Wireless Technologies, Embedded and Intelligent Systems (WITS). 2017.

12. Hashizume, A., P.-S. Yeh, and A. Rosenfeld, A method of detecting the orientation of aligned components. *Pattern Recognition Letters*, 1986. 4(2): p. 125–132.

13. Yue, L., and T. Chew Lim, Improved nearest neighbor based approach to accurate document skew estimation. In: Seventh International Conference on Document Analysis and Recognition, 2003. Proceedings. 2003.

14. Baird, H.S., The skew angle of printed documents. In: Document Image Analysis, O.G. Lawrence and K. Rangachar, Editors, IEEE Computer Society Press. p. 204–208. 1995.

15. Li, S., Q. Shen, and J. Sun, Skew detection using wavelet decomposition and projection profile analysis. *Pattern Recognition Letters*, 2007. 28(5): p. 555–562.

16. Ptak, R., B. Żygadło, and O. Unold, *Projection–Based Text Line Segmentation with a Variable Threshold*, 2017. 27(1): p. 195.

17. Chamchong, R., and C.C. Fung, Text line extraction using adaptive partial projection for palm leaf manuscripts from Thailand. In: International Conference on Frontiers in Handwriting Recognition. 2012.

18. Valy, D., M. Verleysen, and K. Sok, Line segmentation approach for ancient palm leaf manuscripts using competitive learning algorithm. In: 15th International Conference on Frontiers in Handwriting Recognition (ICFHR). 2016.

19. Chamchong, R., and C.C. Fung, Character segmentation from ancient palm leaf manuscripts in Thailand. In: Proceedings of the 2011 Workshop on Historical Document Imaging and Processing, ACM: Beijing, China, USA. p. 140–145.

20. Nicolaou, A., and B. Gatos, Handwritten text line segmentation by shredding text into its lines. In: 10th International Conference on Document Analysis and Recognition. 2009.

21. Peng, G., et al, Text line segmentation using Viterbi algorithm for the palm leaf manuscripts of Dai. In: International Conference on Audio, Language and Image Processing (ICALIP). 2016.

22. Zhang, X., and C.L. Tan, Text line segmentation for handwritten documents using constrained seam carving. In: 14th International Conference on Frontiers in Handwriting Recognition. 2014.

23. Surinta, O., et al., A path planning for line segmentation of handwritten documents. In: 14th International Conference on Frontiers in Handwriting Recognition. 2014.

24. Arvanitopoulos, N., and S. Süsstrunk, Seam carving for text line extraction on color and grayscale historical manuscripts. In: 14th International Conference on Frontiers in Handwriting Recognition. 2014.

25. Valy, D., M. Verleysen, and K. Sok, Line segmentation for grayscale text images of Khmer palm leaf manuscripts. In: Seventh International Conference on Image Processing Theory, Tools and Applications (IPTA). 2017.

26. Hart, P.E., N.J. Nilsson, B. Raphael, and A. Formal, Basis for the heuristic determination of minimum cost paths *IEEE Transactions on Systems Science and Cybernetics*, 1968. 4(2): p. 100–107.

27. Casey, R.G., and E. Lecolinet, A survey of methods and strategies in character segmentation. *IEEE Transactions on Pattern Analysis and Machine Intelligence*, 1996. 18(7): p. 690–706.

28. Shi, Z., and V. Govindaraju, Segmentation and recognition of connected handwritten numeral strings. *Pattern Recognition*, 1997. 30(9): p. 1501–1504.

29. Elnagar, A., and R. Alhajj, Segmentation of connected handwritten numeral strings. *Pattern Recognition*, 2003. 36(3): p. 625–634.

30. Pal, U., A. Belaïd, and C. Choisy, Touching numeral segmentation using water reservoir concept. *Pattern Recognition Letters*, 2003. 24(1): p. 261–272.

31. Pravesjit, S., and A. Thammano, Segmentation of historical lanna handwritten manuscripts In: 6th IEEE International Conference Intelligent Systems. 2012.

32. Ribas, F.C., L.S. Oliveira, A.S. Britto, and R. Sabourin, Handwritten digit segmentation: A comparative study. *International Journal on Document Analysis and Recognition (IJDAR)*, 2013. 16(2): p. 127–137.

33. Elagouni, K., et al, Combining multi-scale character recognition and linguistic knowledge for natural scene text OCR. In: 10th IAPR International Workshop on Document Analysis Systems. 2012.

34. Fujisawa, H., Y. Nakano, and K. Kurino, Segmentation methods for character recognition: from segmentation to document structure analysis. *Proceedings of the IEEE*, 1992. 80(7): p. 1079–1092.

35. Yi-Kai, C., and W. Jhing-Fa, Segmentation of single- or multiple-touching handwritten numeral string using background and foreground analysis. *IEEE Transactions on Pattern Analysis and Machine Intelligence*, 2000. 22(11): p. 1304–1317.

36. Oliveira, L.S., R. Sabourin, F. Bortolozzi, and C.Y. Suen, Automatic recognition of handwritten numerical strings: A recognition and verification strategy. *IEEE Transactions on Pattern Analysis and Machine Intelligence*, 2002. 24(11): p. 1438–1454.

37. Ji, J., L. Peng, and B. Li, Graph model optimization based historical Chinese character segmentation method. In: 11th IAPR International Workshop on Document Analysis Systems. 2014.

38. Chatchinarat, A., Thai handwritten segmentation using proportional invariant recognition technique. In: International Conference on Future Computer and Communication. 2009.

39. Inkeaw, P., et al., Recognition-based character segmentation for multi-level writing style. *International Journal on Document Analysis and Recognition (IJDAR)*, 2018. 21(1): p. 21–39.

40. Dalal, N., and B. Triggs, Histograms of oriented gradients for human detection. In: IEEE Computer Society Conference on Computer Vision and Pattern Recognition (CVPR'05). 2005.

41. Surinta, O., M.F. Karaaba, L.R.B. Schomaker, and M.A. Wiering, Recognition of handwritten characters using local gradient feature descriptors. *Engineering Applications of Artificial Intelligence*, 2015. 45: p. 405–414.

42. Sayyouri, M., A. Hmimid, and H. Qjidaa, A fast computation of charlier moments for binary and gray-scale images. In: Information Science and Technology (CIST), 2012 Colloquium in. 2012.

43. Sayyouri, M., A. Hmimid, H. Qjidaa, and A. Fast, A fast computation of novel set of meixner invariant moments for image analysis. *Circuits, Systems, and Signal Processing*, 2015. 34(3): p. 875–900.

44. Liu, J.G., H.F. Li, F.H.Y. Chan, and F.K. Lam, Fast discrete *cosine* transform *via* computation of moments. *Journal of VLSI Signal Processing Systems for Signal, Image and Video Technology*, 1998. 19(3): p. 257–268.

45. Zhou, J., et al *Image Analysis by Discrete Orthogonal Hahn Moments*. Berlin, Heidelberg: Springer Berlin Heidelberg. 2005.

46. Pew-Thian, Y., R. Paramesran, and O. Seng-Huat, Image analysis by Krawtchouk moments. *Image Processing, IEEE Transactions on*, 2003. 12(11): p. 1307–1377.

47. Mukundan, R., S.H. Ong, and P.A. Lee, Image analysis by Tchebichef moments. *IEEE Transactions on Image Processing : A Publication of the IEEE Signal Processing Society*, 2001. 10(9).

48. Hosny, K.M., Image representation using accurate orthogonal Gegenbauer moments. *Pattern Recognition Letters*, 2011. 32(6): p. 795–804.

49. Gonzalez, R.C., and R.E. Woods, *Digital Image Processing*. 3rd ed. 2006, Prentice-Hall, Inc.

50. Santosh, K.C., Character recognition based on DTW-radon. In: International Conference on Document Analysis and Recognition. 2011.

51. Santosh, K.C., and L. Wendling, Character recognition based on non-linear multi-projection profiles measure. *Frontiers of Computer Science*, 2015. 9(5): p. 678–690.

52. Jain, J., et al., *Modified Chain Code Histogram Feature for Handwritten Character Recognition*. Berlin, Heidelberg: Springer Berlin Heidelberg. 2012.

53. Santosh, K.C., and E. Iwata, Stroke-based cursive character recognition. In: Advances in Character Recognition, X. Ding, Editor, IntechOpen. p. 175–192. 2012.

54. Inkeaw, P., et al., Recognition of handwritten Lanna Dhamma characters using a set of optimally designed moment features. *International Journal on Document Analysis and Recognition (IJDAR)*, 2017. 20(4): p. 259–274.

55. Shridhar, M., and A. Badreldin, High accuracy character recognition algorithm using Fourier and topological descriptors. *Pattern Recognition*, 1984. 17(5): p. 515–524.

56. Wu, T., et al., An improved descriptor for Chinese character recognition. In: Third International Symposium on Intelligent Information Technology Application. 2009.

57. Lecun, Y., L. Bottou, Y. Bengio, and P. Haffner, Gradient-based learning applied to document recognition. *Proceedings of the IEEE*, 1998. 86(11): p. 2278–2324.

58. Dae-Gun, K., S. Song, K. Kang, and S. Han, Convolutional neural networks for character-level classification. *IEIE Transactions on SMART Processing & Computing*, 2017. 6(1): p. 53–59.

59. Bai, J., et al., Image character recognition using deep convolutional neural network learned from different languages. In: IEEE International Conference on Image Processing (ICIP). 2014.

60. Hochreiter, S., and J. Schmidhuber, Long short-term memory. *Neural Computation*, 1997. 9(8): p. 1735–1780.

61. Graves, A., et al., Connectionist temporal classification: Labelling unsegmented sequence data with recurrent neural networks. In: Proceedings of the 23rd International Conference on Machine Learning, ACM: Pittsburgh, Pennsylvania, USA. p. 369–376. 2006.

62. Ahmed, S.B., et al., Evaluation of cursive and non-cursive scripts using recurrent neural networks. *Neural Computing and Applications*, 2016. 27(3): p. 603–613.

63. Sarraf, S., French word recognition through a quick survey on recurrent neural networks using long-short term memory RNN-LSTM. *Arxiv e-Prints*, 2018.
64. Inkeaw, P., et al., *Rule-Based Page Segmentation for Palm Leaf Manuscript on Color Image*. Cham: Springer International Publishing. 2016.
65. Burger, W., and M. Burge, *Digital Image Processing : An Algorithmic Introduction Using Java*. London: Springer Publishing Company. 2016.

# *Index*

Printed and bound by CPI Group (UK) Ltd, Croydon, CR0 4YY

23/10/2024

01778223-0020